An Ethical Guide to Cyber Anonymity

Concepts, tools, and techniques to protect your anonymity
from criminals, unethical hackers, and governments

Kushantha Gunawardana

BIRMINGHAM—MUMBAI

An Ethical Guide to Cyber Anonymity

Group Product Manager: Mohd Riyan Khan

Publishing Product Manager: Mohd Riyan Khan

Senior Editor: Athikho Sapuni Rishana

Technical Editor: Shruthi Shetty

Copy Editor: Safis Editing

Project Coordinator: Deeksha Thakkar

Proofreader: Safis Editing

Indexer: Sejal Dsilva

Production Designer: Ponraj Dhandapani

Marketing Coordinator: Ankita Bhonsle

First published: December 2022

Production reference: 1171122

Published by Packt Publishing Ltd.
Livery Place
35 Livery Street
Birmingham
B3 2PB, UK.

ISBN 978-1-80181-021-0

www.packt.com

I would like to dedicate this book to my beloved mom Deeliya and dad Chandra for the sacrifices they have made, my mother-in-law Chithra for all her dedication and support rendered, and especially to my courageous wife Amali, my brilliant boys Rivindu and Risindu, and amazing daughter Asekha for their love and support, which helped me to use our very precious yet limited family time to make my dream a reality!

I know my mom, dad, and my mother-in-law are very proud of this, but unfortunately, they wouldn't really know what this book is all about, as they are from a wonderful generation that was totally disconnected from the cyber world!

Also, I would like to dedicate this book to all my students around the world, the attendees to my training sessions, who always encouraged me to write a book, and all my friends and colleagues for believing in me!

Kushantha Gunawardana

Contributors

About the author

Kushantha Gunawardana is an internationally recognized cybersecurity trainer, security consultant, YouTuber, blogger, and public speaker who won the EC-Council Instructor Circle of Excellence Global Award in 2021. In his 20 years of vast exposure to cybersecurity, he has contributed to many overseas projects and trained thousands of university students, professionals, government employees, and military, and police officers in cybersecurity and forensics in 25 countries.

He holds a master's degree in IT specializing in networking with distinction from the University of Colombo, was awarded the JICA Award 2011 (Gold Medal), led the team who won the National Hacking Challenge in Sri Lanka in 2013, and is reading for a Ph.D. in cyber forensics.

I want to thank the people who have been close to me, believing in me, and supporting me! Finally, the dedicated team at Packt publication, who pushed me throughout the journey to get this amazing outcome!

About the reviewer

Ahmad Muammar WK currently works as an offensive security director at Seclab Indonesia. He holds **Offensive Security Certified Professional (OSCP)**, **Offensive Security Certified Expert (OSCE)**, and **eLearnSecurity Mobile Application Penetration Tester (eMAPT)** certifications. He is the founder of ECHO, one of the oldest Indonesian IT security communities, and also a founder of IDSECCONF, the biggest annual security conference in Indonesia. He also reviewed *Kali Linux Cookbook*, by Willie L. Pritchett and David De Smet, Packt Publishing, and *Kali Linux Network Scanning Cookbook*, by Justin Hutchens, Packt Publishing.

Table of Contents

Part 2: Methods and Artifacts That Attackers and Competitors Can Collect from You

3

Ways That Attackers Reveal the Privacy of Individuals and Companies

4

Techniques that Attackers Use to Compromise Privacy   75

5

6

Part 3: Concepts and Maintaining Cyber Anonymity

7

10

Proxy Chains and Anonymizers   249

Index   289

Other Books You May Enjoy   302

Preface

As the world becomes more connected through the web, new data collection innovations have opened ways to exploit privacy. Your actions on the web are being tracked, information is being stored, and your identity could be stolen. However, there are ways to use the web without risking your privacy. This book will take you on a journey to become invisible and anonymous while using the web.

You will start the book by understanding what anonymity is and why it is important. Understanding the objective of cyber anonymity, you will learn to maintain anonymity and perform tasks without disclosing your information. Then, you'll learn how to configure tools and understand the architectural components of the cybereconomy. Finally, you will learn to be safe during intentional and unintentional internet access by taking the relevant precautions.

By the end of this book, you will be able to work with the internet and internet-connected devices safely by maintaining cyber anonymity.

Who this book is for

This book is targeted at journalists, security researchers, ethical hackers, and anyone who wishes to stay anonymous while using the web. This book is also for parents who wish to keep their kid's identities anonymous on the web.

What this book covers

Chapter 1, Understanding Sensitive Information, will provide you with an understanding of sensitive information and what privacy and cyber anonymity are – this is the introductory chapter for the book.

Chapter 2, Ways That Attackers Use Your Data, will explain why attackers are interested in your privacy and the ways that attackers use stolen data after data breaches. Often, attackers use stolen data to commit more crimes or as entry points to attack other companies.

Chapter 3, Ways That Attackers Reveal the Privacy of Individuals and Companies, will detail how attackers reveal the privacy of individuals and companies.

Chapter 4, Techniques that Attackers Use to Compromise Privacy, will describe what types of techniques attackers will use to compromise your privacy.

Chapter 5, Tools and Techniques That Attackers Use, will uncover what different tools attackers use to compromise privacy. This will improve the reader's knowledge of the different kinds of tools that attackers can use.

Chapter 6, Artifacts that Attackers Can Collect from You, will outline what type of data attackers are interested in collecting from companies and individuals.

Chapter 7, Introduction to Cyber Anonymity, will introduce you to cyber anonymity and cover the basics of the layers of cyber anonymity.

Chapter 8, Understanding the Scope of Access, will provide information on how to set up cyber anonymity and requirements to maintain cyber anonymity.

Chapter 9, Avoiding Behavior Tracking Applications and Browsers, will explain how to maintain cyber anonymity and the areas and techniques that we can use to maintain cyber anonymity.

Chapter 10, Proxy Chains and Anonymizers, will explain the tools and techniques that can be used to maintain cyber anonymity.

Download the color images

We also provide a PDF file that has color images of the screenshots and diagrams used in this book. You can download it here: `https://packt.link/PmYh6`.

Conventions used

There are a number of text conventions used throughout this book.

`Code in text`: Indicates code words in text, database table names, folder names, filenames, file extensions, pathnames, dummy URLs, user input, and Twitter handles. Here is an example: "Attackers can dump this file using the `fgdump`, `samdump`, and `pwddump` tools."

Bold: Indicates a new term, an important word, or words that you see onscreen. For instance, words in menus or dialog boxes appear in **bold**. Here is an example: "On the **Security** tab, you can see the list of users including built-in and inherited users that have access to the given object."

> **Tips or important notes**
> Appear like this.

Get in touch

Feedback from our readers is always welcome.

General feedback: If you have questions about any aspect of this book, email us at customercare@ packtpub.com and mention the book title in the subject of your message.

Errata: Although we have taken every care to ensure the accuracy of our content, mistakes do happen. If you have found a mistake in this book, we would be grateful if you would report this to us. Please visit www.packtpub.com/support/errata and fill in the form.

Piracy: If you come across any illegal copies of our works in any form on the internet, we would be grateful if you would provide us with the location address or website name. Please contact us at copyright@packt.com with a link to the material.

If you are interested in becoming an author: If there is a topic that you have expertise in and you are interested in either writing or contributing to a book, please visit authors.packtpub.com.

Share Your Thoughts

Once you've read *An Ethical Guide to Cyber Anonymity*, we'd love to hear your thoughts! Scan the QR code below to go straight to the Amazon review page for this book and share your feedback.

https://packt.link/r/1801810214

Your review is important to us and the tech community and will help us make sure we're delivering excellent quality content.

Download a free PDF copy of this book

Thanks for purchasing this book!

Do you like to read on the go but are unable to carry your print books everywhere?

Is your eBook purchase not compatible with the device of your choice?

Don't worry, now with every Packt book you get a DRM-free PDF version of that book at no cost.

Read anywhere, any place, on any device. Search, copy, and paste code from your favorite technical books directly into your application.

The perks don't stop there, you can get exclusive access to discounts, newsletters, and great free content in your inbox daily

Follow these simple steps to get the benefits:

1. Scan the QR code or visit the link below

https://packt.link/free-ebook/978-1-80181-021-0

2. Submit your proof of purchase
3. That's it! We'll send your free PDF and other benefits to your email directly

Part 1: The Basics of Privacy and Cyber Anonymity

Upon completion of this part, you will understand the basics of privacy and cyber anonymity.

This part comprises the following chapters:

- *Chapter 1, Understanding Sensitive Information*
- *Chapter 2, Ways That Attackers Use Your Data*

1
Understanding Sensitive Information

Before we start learning about the concept of cyber anonymity, it's important to understand the level of sensitivity of information. In today's world, information is power. If you look at the wealthiest companies in the world, all of them are related to information. Typically, we think illegal activities, including dealing drugs, selling weapons, and smuggling, generate lots of money and power, or "create kingdoms." But the reality is, information has power exceeding all of these underground activities.

The world's top wealthiest companies and individuals have gained this status by managing information. Typically, data is in the raw form of facts and statistics. This can be used for reference or analysis. Once properly analyzed, data becomes information. Information is generally processed data that gives us meaningful context that can be used for decision-making. That's why information has become power as information is processed, structured, and organized data that enables powerful decision-making.

As an example, let's take an advertising campaign that utilizes TV or social media advertisements. With TV broadcasting, it will broadcast to millions of people, but the target customer engagement would be a very low percentage. If social media is used, we could select the precisely interested or prospective users to advertise to. So, the impact will be very high. This is more powerful than we think. If you select the exact audience that you want the advertisement to reach, selecting attributes such as age group, gender, and geography, it will be more effective. Not only this, but nowadays, social media even has data about users' genuine likes and dislikes. If we use social media, the advertisement will be delivered to prospective users. This is also known as direct marketing.

This chapter will cover the following main topics:

- The categorization of information
- Different forms of sensitive information
- Raw data can create sensitive information
- Privacy in cyberspace
- Cyber anonymity

The categorization of information

Information can be further classified and categorized depending on the sensitivity. As an example, in today's world, mobile phones have also become information repositories. Everyone's mobile phone has a large amount of information that they have stored intentionally or unintentionally. Nowadays, information can be in different forms, not just text or numbers. It can be in the form of documents, images, videos, and so on. Some information is stored by users on their mobile phones intentionally. Users of mobile phones are aware that this information is stored. But there is also another set of information stored in phones without users' knowledge.

People often confuse personal and sensitive information. Collecting, storing, using, or disclosing sensitive information is protected under different lawsuits around the world. A famous one is GDPR, passed by the European Union in 2016 and enforced in May 2018. These legal concerns are very strict on sensitive information. The reason behind this is disclosing sensitive information can have an irreversible effect on someone's life. Let's look at the difference between personal information and sensitive information:

- **Personal information**: Personal information refers to any information about an individual or a person that makes them distinguishable or identifiable. Under the law, even if the information given is not accurate, it is still considered personal information.

  Personal information includes an individual's name, address, contact information, date of birth, email address, and bank information.

- **Sensitive information**: Unlike personal information, sensitive information has a direct impact on the individual if disclosed. Sensitive information is a subcategory of personal information in a broader sense. Sensitive information may have a direct impact on or harm an individual if it is not handled properly.

Sensitive information includes an individual's criminal record, health records, biometric information, sexual orientation, or membership in a trade union. If disclosed, the result may be discrimination, harassment, or monetary loss for the person to whom the sensitive information pertains.

If you look at the aforementioned information, what we need to understand is most **Personally Identifiable Information** (**PII**) is not confidential to our close relations and friends. Also, nowadays, our close circles have expanded to the global level with social media. Most social media users overexpose their own or other people's personal information, either intentionally or unintentionally.

Different forms of sensitive information

Most users aren't aware that when they access a web application or a website, it can collect some of their information. They just think that they are only accessing information from the web browser, but the reality is web applications can collect a lot more information than users think. To understand this, we can simply access `https://www.deviceinfo.me` (as shown in *Figure 1.1*). This website shows you how much information is collected from your device just by accessing a website. If you access this website with your mobile phone, it will display lots of information, including your mobile phone's type/model, operating system, browser version, IP address, hostname, number of cores, memory, interfaces, and latitude and longitude. This shows that web applications and websites can collect almost all the information about a device.

Figure 1.1 – Information derived about your device

This is a classic example of the data that a simple web application can collect, just by getting a device to access the application, without installing any agent or running a script.

When you look at the data that we have on our devices, mobile phones, or desktops, it can be sorted into a few categories. But not every case will contain PII.

Any form of information that could lead to any type of loss, such as financial, if accessed by a third party can be considered sensitive information.

Sensitive information can take different forms.

Mostly, people think that sensitive information is banking account information, including credit card numbers and social media account information. But a private picture or video clip can be even more sensitive than the preceding listed examples of sensitive information.

If you lose your credit card, the maximum damage that can occur is the credit limit of the stolen credit card is reached. But if someone accesses a private picture or video clip of yours, it can create more damage that might not be reversible.

Sometimes, we disclose sensitive information unintentionally. Let's look at the following photo:

Figure 1.2 – Photo of a car, carrying sensitive information

This photo was taken by someone trying to sell their car. As a precaution, the seller has even masked the numberplate to reduce the information this photo discloses. Even if you inspect this photo closely, you might not find any interesting information. But although the seller has masked part of the vehicle identification number, there is still a lot of information given away with this photo without their knowledge. This information is known as metadata. Metadata can be defined as data about data.

We usually look at the content of a file, but metadata discloses even more information than we are aware of.

Let's look at the metadata of the preceding photo. Let's access `http://metapicz.com/`, upload the image, and see what we can find. This site can acquire meta information on an image. It extracts information including the camera make, model, exposure, and aperture of the device.

Figure 1.3 – Derived information from a picture

The preceding screenshot shows the make of the device that captured the photo of the car, as well as the model and exposure. Exposure refers to the amount of light that comes in while you are pressing the capture button of a camera. Aperture refers to the opening of the lens of the camera to allow light through and focal length is the distance between the lens and the image sensor. So, if someone analyzes an image, they can get a massive amount of information, even about the lighting conditions of the environment during the time of capturing this image.

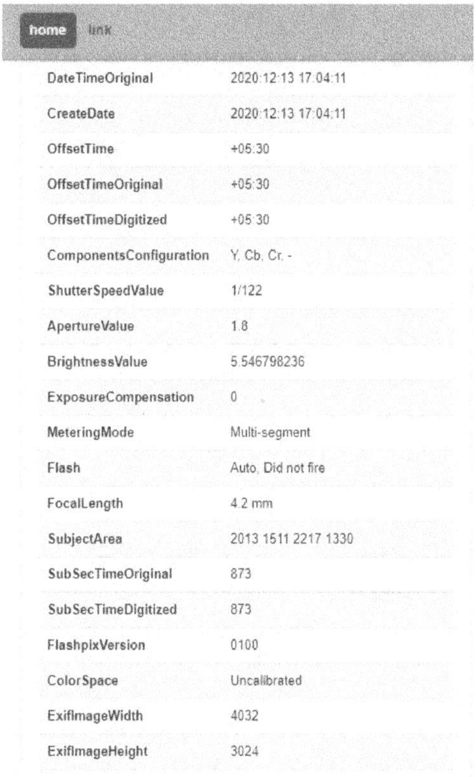

home	link	
DateTimeOriginal	2020:12:13 17:04:11
CreateDate	2020:12:13 17:04:11
OffsetTime	+05:30
OffsetTimeOriginal	+05:30
OffsetTimeDigitized	+05:30
ComponentsConfiguration	Y, Cb, Cr, -
ShutterSpeedValue	1/122
ApertureValue	1.8
BrightnessValue	5.546798236
ExposureCompensation	0
MeteringMode	Multi-segment
Flash	Auto, Did not fire
FocalLength	4.2 mm
SubjectArea	2013 1511 2217 1330
SubSecTimeOriginal	873
SubSecTimeDigitized	873
FlashpixVersion	0100
ColorSpace	Uncalibrated
ExifImageWidth	4032
ExifImageHeight	3024

Figure 1.4 – More information from the image metadata

The preceding screenshot shows information related to the time the photo was taken, such as the created date and offset. Typically, offset refers to the time zone. According to this screenshot, the offset is **+05.30**, which refers to GMT +5:30, which is Asia/Colombo time, specifically, Kolkata. By analyzing this, we now know the region in which the photo was taken.

home	link

FocalLengthIn35mmFormat	26 mm
SceneCaptureType	Standard
LensInfo	4.25-6mm f/1.8-2.4
LensMake	Apple
LensModel	iPhone XS Max back dual camera 4.25mm f/1.8
CompositeImage	General Composite Image
GPSLatitudeRef	North
GPSLatitude	6.877256
GPSLongitudeRef	East
GPSLongitude	79.937386
GPSAltitudeRef	Above Sea Level
GPSAltitude	18.66111978 m
GPSSpeedRef	km/h
GPSSpeed	0.7953313592
GPSImgDirectionRef	True North
GPSImgDirection	289.1087645
GPSDestBearingRef	True North
GPSDestBearing	289.1087645
GPSHPositioningError	41.62442334 m

Figure 1.5 – Lens and GPS information derived from the image

Personally identifiable data or information is anything that discloses information about you, including your name, address, telephone number, or social media identity, photos with contents that identify you, and even metadata. Also, your email address or IP can be treated as PII. We should be able to control our privacy and decide when, how, and to what extent our PII is revealed.

This is also known as **data privacy**. There are many initiatives and acts around the globe that relate to data privacy, but data privacy can be violated at various levels. Most of the devices we use today compromise our privacy even without us being aware. We will take Android as an example. Whenever you use an Android phone, it collects a lot of information about you as we usually connect our Gmail account to get the full functionality of the Android device. Once you have connected your Gmail account to the Android device, it will start collecting your information.

If you want to see what information about you Google is collecting, access `https://myactivity.google.com/` and log in using your Google account that is connected to your device. You will be amazed to see how much information Google collects, including your web and app activity, your location history, and your YouTube watch history, that is, all the videos you watch and search for on YouTube.

If you go to the location history and click the **Manage Activity** link, you will see how much data your device has uploaded to Google.

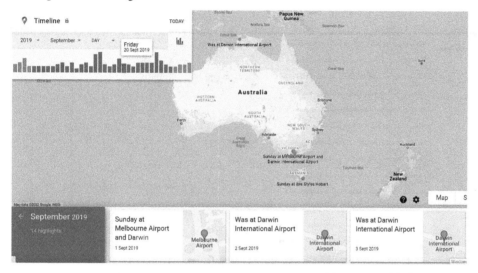

Figure 1.6 – Activity recorded in Google

If you select any of the dates, it will show you all your movements, including the method of commute, very accurate information about your walk, and even photos that you took taken during the journey using your device's camera.

Figure 1.7 – Detailed information captured by Google

Even if you disable data connectivity on your Android device while traveling, the device will still collect all this information and upload it to Google whenever you connect your device to the internet later. This doesn't just occur on the Android operating system; all devices do this – even your iPhone or Windows device.

If you want to check what information is being stored on your Windows device, press the Windows key + *I* to access **Windows Settings | Privacy | Diagnostics & feedback | Open Diagnostic Data Viewer**.

Figure 1.8 – Telemetric data shared with Microsoft by your device

This shows what your device is sharing with Microsoft. Not only operating systems but also applications collect our information. You may have noticed that many applications that you install on your device request access to your location, photos, and other sources of information, even if the app doesn't need to use this information. As an example, if you install a flashlight app and it requests access to your location, that is suspicious.

The reality is most operating systems, applications, devices, and even manual systems collect this information. Sometimes, disclosing personal information can be dangerous. There was an incident reported in India recently related to this. A **Business Process Outsourced** (**BPO**) company that provides services to overseas companies from India was advertising a vacancy. Many people came for the interview. At the security post, there was an open register on which each candidate had to fill in information, including their name, address, telephone number, and email address. (This is common in many Asian countries.) There was a woman who attended the interview that got a call for a second interview at a different location in the evening. It was a bit suspicious, but the BPO company operates 24x7, so this wasn't too odd of a scenario as during the first interview, they informed candidates that the job would be on a shift basis. The woman went for the second interview but never came back. Later investigations found that the call for the second interview was fake; her information was collected by someone who came to the same interview and as everyone was filling in an open registry, they were able to access all previous records. This shows how dangerous disclosing personal information can be. But still, I have seen many locations where retail stores do it especially during the 2020 pandemic, as they wanted to trace positive cases of COVID-19 and inform people who had come into contact with those infected people.

Raw data can create sensitive information

There are different ways we disseminate our data knowingly or unknowingly, including participating in different types of surveys. Sometimes, researchers who conduct surveys may not use the data for the intended purpose. Often, data collected as part of open or public surveys is used for different purposes. If a researcher is collecting data, it's important that they disclose the purpose of the data collection, and the data collected cannot and should not be used for any other purpose than that.

The main advantage of having raw data for an attacker is that this raw data can be processed to get PII or sensitive information, which can be used for direct and indirect attacks.

In the previous section, we discussed what PII and sensitive information are. Let's take an example. When you call a bank or service provider, typically, they ask a few questions to verify your identity. The questions they ask are really basic; as we discussed earlier, this might even be information you've shared with your close circle. These questions can include your full name, address, contact number, and email address. (In fact, I still remember the full names of most of my schoolmates as the first thing that is done in the morning at schools in my country is marking the register. Teachers usually read names aloud one by one, and if the student is present, they have to shout, "Present!" Because of this, I still remember most of my classmates' full names and initials, even though we have long names in our culture!)

Why do service providers ask these sorts of questions? Because by collecting a series of information such as this, they can identify that they are communicating with the correct person. This is the principle behind claims-based authentication in federation trusts. Claims, rather than credentials, will be shared between the identity provider and service provider. Claims are typically attributes, and they are treated as raw data.

Another interesting fact is, once someone has collected raw data, they can easily find personal and sensitive information too. For these types of searches, attackers use different tools. One such tool is Social Searcher (`https://www.social-searcher.com/`).

If you want to find more information on someone, so long as you know their first name and last name, you also can start searching for them on social media. The Social Searcher web app is connected to multiple social media APIs and provides information related to the searched name and its respective social media accounts. There are many internet resources and tools like this that can be found in many open source distributions, including Kali, Parrot OS, Security Onion, and Predator. Later chapters will discuss different tools and techniques in detail to understand what type of integration these tools have with collecting information and how can we prevent creating sensitive information.

Privacy in cyberspace

Every country has its own jurisdiction system and laws. Typically, respective laws are applicable within the country. Even within federal governments, sometimes different states will have different laws and acts. The reason is if any incident takes place, the law enforcement bodies of respective areas or state stake the required actions.

Cyberspace works completely differently, though, as in most incidents, the perpetrator connects remotely over the internet with the target. Most of the time, the attacker is located in a different state or country than the target. The internet is an unregulated space and no one has direct ownership. Every time we access the internet, we need to remember that we are connected to an unregulated space and we need to look after our own security as the internet cannot be completely governed due to its architecture and very nature.

Whenever we are connected to cyberspace using any type of device, we are risking breaches of our privacy. In reality, we compromise our privacy in different layers. When we connect to the internet, we use different kinds of devices. It can be a mobile device, laptop, desktop, and so on. The first layer is the device that we connect to the internet as it stores lots of information. Then, we have the application that we use to access and surf the internet, typically a browser. The browser also keeps lots of information. Then, the device must be connected to the internet using some sort of media; this can be a wired or wireless connection. Whatever connection we use, there is a possibility that the network is collecting information on us. This is known as **network capturing**.

The next layer is the devices to which the network is connected. This includes Wi-Fi routers, switches, and firewalls that are connected to the network, and they also collect information. The network connection is then connected to the **Internet Service Provider** (**ISP**), which collects different kinds of information about the connection. If you are accessing a particular website, the host web server collects information about the connection. This information includes the timestamp, your public IP address, the type of browser being used, and the operating system.

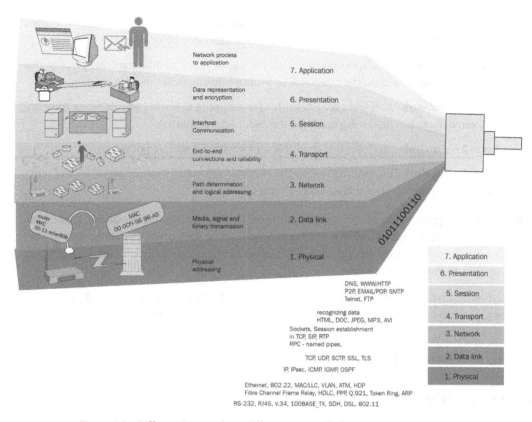

Figure 1.9 – Different layers where different types of information pass through

When you look at these layers, even if an attacker is not involved, there are multiple layers where information is being collected about your connection. As we discussed earlier, this includes personal and sensitive information about you.

This gives an understanding of the different layers between the web application and the device. Importantly, if any of the layers are compromised by attackers, it will become more crucial as then attackers have control and access to these layers. They can even intercept communications and acquire credentials if the protocols of the communication used are weak.

When you perform any activity on the internet, or within the network using an application, data goes through these layers when communicating with other entities. As an example, when you draft an email using an email client such as Microsoft Outlook, data is created in the application layer and all the other layers are responsible for different tasks:

- **Layer 7 – application**: This layer is where the users are directly interacting with the device using an application such as a browser.

- **Layer 6 – presentation**: This layer is preparing source and destination devices to communicate with each other. Encryption and decryption take place in this layer.

- **Layer 5 – session**: This layer helps to establish, manage, and terminate the connection between the source and destination devices. Communication channels are referred to as sessions.

- **Layer 4 – transport**: Transporting data from the source to the destination takes place in this layer. If the dataset is large, then data will be broken into pieces in this layer.

- **Layer 3 – network**: This layer is responsible for mapping the best paths for data traversal between devices and delivering messages through nodes or routers.

- **Layer 2 – data link**: This layer is responsible for switching connected devices.

- **Layer 1 – physical**: This layer represents the physical connectivity, including cables and other mediums responsible for sending data as frames.

When thinking of privacy, you need to concentrate on all the layers, including the device, application, network, communication, and web servers. For example, if you are using a shared device and save passwords on the web browser, your private information can be stolen easily, as there are many free tools out there to make life more easy.

If you access `https://www.nirsoft.net/utils/web_browser_password.html`, you can download web browser password viewers, which can retrieve stored passwords easily.

The same site also provides you with a range of free tools that can be used with the **Graphical User Interface (GUI)** or scripts (command-line tools) to automate the process.

Cyber anonymity

We have discussed how our privacy can be compromised and different levels of privacy.

Cyber anonymity is the state of being unknown. With cyber anonymity, the activities performed in cyberspace will remain, but the state will be unknown. As an example, if an attacker performs an attack anonymously, the attack will still be effective but the attacker's identity will be unknown. Being completely anonymous is a complex process as there are multiple layers of collecting information, as explained earlier.

If we look at the same set of layers that we discussed, to be anonymous in cyberspace, we need to concentrate on each layer. The main idea here is for the attacker to eliminate all traces of themselves as if even a single amount of information is left, they can be identified. That's how many anonymous groups have been traced, in some cases after many years of research.

There was one case related to the world-famous Silk Road, an anonymous marketplace on the dark web mostly selling drugs to over 100,000 buyers around the world. Later, the **Federal Bureau of Investigation (FBI)** seized the site. With the site, the United States government seized over 1 billion US dollars' worth of Bitcoin connected to Silk Road. Even though the main actors behind Silk Road were arrested, the administrators of the site started Silk Road 2, but that was also seized by the US government. However, the site was completely anonymous for a few years until the FBI traced and shut it down. According to the media, the infamous Dread Pirate Roberts, the pseudonym of Ross Ulbricht, the founder of Silk Road, was taken down because of a misconfigured server. This server was used to maintain the cyber anonymity of Silk Road, but due to a single misconfiguration, it uncovered the real IPs of some requests instead of them being anonymous. As a result, the FBI was able to track down the communication and traced the perpetrator using the IP.

This is a classic example to illustrate how even though efforts were made to remain anonymous on all layers, a small mistake revealed their whereabouts. This is why it is stated that cyber anonymity is a complicated process that involves various technologies. Also, it requires concentrating on all the layers to be completely anonymous. There are many common technologies, including **Virtual Private Networks (VPNs)**, proxy servers, censorship circumvention tools, and chain proxies, that help with maintaining cyber anonymity, which will be discussed in upcoming chapters.

Typically, all operating systems, applications, and appliances are designed to keep different types of information in the form of logs to maintain accountability and to be able to help with troubleshooting. This information can be volatile or static. Volatile information will be available until the next reboot or shutdown of the system in memory. Forensic and memory-capturing tools can be used to dump volatile data, which can then be analyzed to find out specific information.

Static data can be found in temporary files, registries, log files, and other locations, depending on the operating system or application. Some information that is available is created by the user activity and some is created as a part of the system process.

If you need to maintain complete anonymity, this information is useful as you need to minimize or prevent the footprints created in different layers. To overcome this challenge, the most used technique is using live boot systems. Most Linux systems provide the flexibility of running a live operating system, using CDs/DVDs, live boot USB drives, or virtual systems directly connected to an ISO file. Some operating systems that have the live boot option available are as follows:

- Kali Linux live boot – penetration testing environment
- Parrot Security or Parrot OS live boot – security testing
- Gentoo – based on FreeBSD
- Predator OS
- Knoppix – based on Debian
- PCLinuxOS – based on Mandrake
- Ubuntu – based on Debian
- Kubuntu – KDE Ubuntu version
- Xubuntu – light Ubuntu version that uses an Xfce desktop environment
- Damn Small Linux – Debian (Knoppix remaster)

- Puppy Linux – Barry Kauler wrote almost everything from scratch
- **Ultimate Boot CD (UBCD)** – diagnostics CD
- openSUSE Live – based on the Jurix distribution
- SystemRescue CD – Linux system on a bootable CD-ROM for repairing your system and your data after a crash
- Feather Linux – Knoppix remaster (based on Debian)
- FreeBSD – derived from BSD
- Fedora – another community-driven Linux distribution
- Linux Mint – an elegant remix based on Ubuntu
- Hiren's BootCD **PE (Preinstallation Environment)** – Windows 10-based live CD with a range of free tools

Once you boot from live boot systems, it reduces or prevents creating logs and temporary files on the actual operating system straight away. Once the live boot system is shut down or rebooted, volatile data and static data are created because your activities are completely removed; when you boot next time, it will be a brand-new operating system. If you require, you always have the option to permanently install most of these operating systems.

Whenever you access the internet, DNS information will be cached in the local system until you manually remove it, the **Time to Live (TTL)** value is reached, or you run an automated tool. When you access any website, the local DNS resolver resolves it and keeps it in the cache until the TTL value becomes 0. When configuring DNS on the domain service provider's portal or DNS server, usually, the TTL values are added.

As an example, by using the `nslookup` command, we can check the TTL value.

Let's use `nslookup` on `microsoft.com`:

```
Command Prompt - powershell

PS C:\Users\kusha> nslookup
Default Server:  192-168-1-1.tpgi.com.au
Address:  192.168.1.1

> set q=any
> microsoft.com
Server:  192-168-1-1.tpgi.com.au
Address:  192.168.1.1

Non-authoritative answer:
microsoft.com   internet address = 104.215.148.63
microsoft.com   internet address = 40.76.4.15
microsoft.com   internet address = 40.112.72.205
microsoft.com   internet address = 40.113.200.201
microsoft.com   internet address = 13.77.161.179
microsoft.com   nameserver = ns1-205.azure-dns.com
microsoft.com   nameserver = ns2-205.azure-dns.net
microsoft.com   nameserver = ns3-205.azure-dns.org
microsoft.com   nameserver = ns4-205.azure-dns.info
microsoft.com
        primary name server = ns1-205.azure-dns.com
        responsible mail addr = azuredns-hostmaster.microsoft.com
        serial  = 1
        refresh = 3600 (1 hour)
        retry   = 300 (5 mins)
        expire  = 2419200 (28 days)
        default TTL = 300 (5 mins)
microsoft.com   MX preference = 10, mail exchanger = microsoft-com.mail.protection.outlook.com
microsoft.com   text =
```

Figure 1.10 – DNS information retrieval with nslookup

This shows the TTL value of microsoft.com is 300 seconds/5 minutes.

If we access the Microsoft website, this DNS entry will be cached in the local cache.

We can check this by executing ipconfig /displaydns on Windows Command Prompt.

```
PS C:\Users\kusha> ipconfig /displaydns

Windows IP Configuration

    login.microsoftonline.com
    ----------------------------------------
    Record Name . . . . . : login.microsoftonline.com
    Record Type . . . . . : 5
    Time To Live  . . . . : 215
    Data Length . . . . . : 8
    Section . . . . . . . : Answer
    CNAME Record  . . . . : ak.privatelink.msidentity.com
```

Figure 1.11 – Information retrieved by ipconfig/displaydns

If you are using PowerShell, you can use the `Get-DnsClientCache` cmdlet to get a similar result.

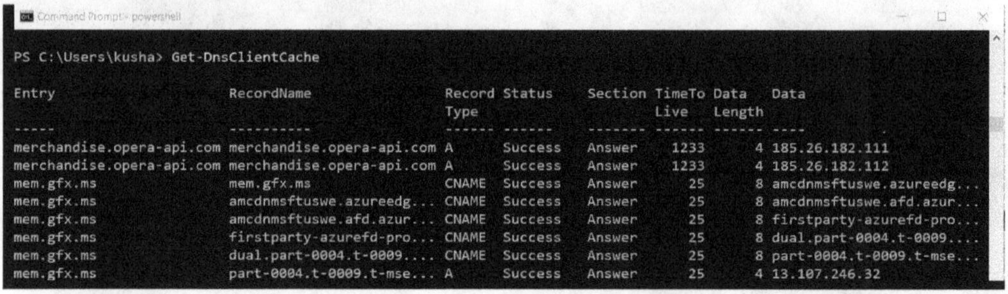

Figure 1.12 – Information retrieved by Get-DnsClientCache

This information is categorized as volatile information. However, until your next reboot or shutdown, these entries will be there if the TTL value has not reached 0.

If you execute the preceding command a few times, with some intervals, you will realize every time you run it, the TTL value of the result is always less than the previous TTL value. When the TTL value becomes 0, the entry will be automatically removed. This is how DNS has been designed, to provide optimum performance during the runtime and when you change the DNS entry. That's the reason why when you change the DNS entry, it can take up to 48 hours to completely replicate the DNS as some clients might still have resolved IPs from DNS entries in their cache.

This is not just the case on the local cache; if you have DNS servers in the infrastructure, these DNS servers also cache the resolved DNS entries for later use.

Summary

This chapter focused on five core areas to provide a clear foundation for cyber anonymity. We learned how to identify sensitive information and categorize and classify it. We also learned about the ways that an attacker can retrieve sensitive information from raw data. We also discussed privacy concerns in cyberspace and areas to look at when it comes to cyber anonymity.

In the next chapter, you will learn the reasons why attackers are interested in breaching your privacy and how attackers use stolen data for their benefit.

2
Ways That Attackers Use Your Data

The constant data breaches reported around the globe have become common news in today's world. Many hackers and hacker communities keep on compromising systems to steal data. These data breaches range from personal-level data breaches to enterprise-level, damaging companies' economies and reputations. Often, attackers use stolen data to commit more crimes or as entry points to attack other companies.

When it comes to concentrating on data privacy in cyberspace, it is important to understand the ways that attackers use stolen data after data breaches.

In this chapter, we will be focusing on the following:

- Impersonation and identity theft
- Technical, procedural, and physical access
- Technical controls
- Procedural controls
- Physical controls
- Creating vulnerabilities to compromise systems
- Increasing the attack surface using sensitive data

Once privacy is compromised, attackers will have your data and information. When attackers have access to personal or sensitive information, they can utilize it for different purposes. Understanding the different ways that attackers use your information will help you to take precautions and even if there is a breach, this will help you to manage it effectively.

Primarily, attackers can use stolen data for impersonation and identity theft.

Impersonation and identity theft

Impersonation is when the attacker uses stolen data to pretend to be someone else. This information can be collected from social media or any type of privacy compromise. Then, the attacker can pretend to be you for different purposes.

Identity theft is when an attacker uses stolen data to access your bank accounts and create fake IDs and even passports. Also, attackers can use stolen personal information to get financial benefits such as car loans, credit cards, and checks without your knowledge or any type of consent. When they receive financial benefits, you will be naturally liable for them. The information stolen can be different from country to country and region to region. For example, in the United States, the most targeted piece of personal information is your social security number. Attackers use this to get financial benefits and even to release criminals on bail using stolen social security numbers, and eventually, the real owner will be liable if the criminals do not show up in the courts. Then, the real owner will be blacklisted financially and legally. They will not be able to get financial or any other benefits as they will have bad credit records.

Both impersonation and identity theft are illegal and can be prosecuted on different levels depending on the damage and the impact. Identity theft impacts millions of people annually around the world, and this can create a huge financial impact on the victims.

Impersonation and identity theft are not always a result of cyberattacks. They can be a result of reckless handling of your paper-based information. As an example, if you dump your bank account statements or credit card statements in the bin, they can be used by attackers to retrieve your information. This is often known as dumpster diving. Dumpster diving is an act of searching trash bins to obtain useful information. This information will later be used by attackers for impersonation and identity theft. Attackers can even use workstations and laptops that have been disposed of to obtain useful information by recovering data. Useful information that attackers would be interested in can include the following:

- Full name/address/email address
- Phone numbers: There are many tools available to extract phone numbers from an email, text, or the web.
- Credentials that are sent by postal mail or emails, particularly web hosting account information sent to your email address.
- Bank and financial statements.
- Important documents with personal and sensitive data.
- Health records.
- Confidential correspondence including trade secrets or business secrets.
- Employee records.
- Insurance and financial information.

The following screenshot shows stolen documents that are on the market on the dark web:

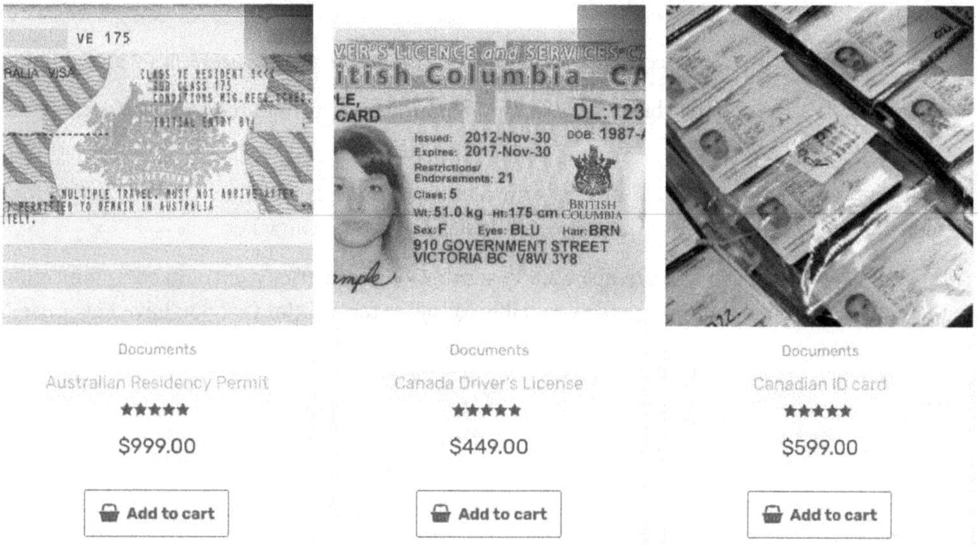

Figure 2.1 – Stolen documents available to buy on the dark web

Mostly, identity theft focuses on financial gains, but there can be other types of motivation by attackers. There are a few common scenarios in which attackers commit identity theft:

- Compromising mail accounts: When an attacker compromises mail accounts, they will hunt for useful information including unencrypted credit card information, tax-related communications, social media account-related emails, correspondence with banks, and communication with other financial services such as PayPal.

- Phishing and spear phishing: Attackers will trick users into submitting their sensitive information on fake websites that look and feel the same as authentic websites.

- Data breaches: This is when users or businesses expose sensitive information due to poor security and practices or because of hacking. Often, users or companies unknowingly share sensitive information in vacancy advertisements and professional communities such as LinkedIn.

Once attackers compromise your private or sensitive information, there are common ways attackers utilize this information:

- Using stolen information, they can apply for credit cards or loans.

- Attackers can get tax refunds transferred to their accounts.

- Attackers can redirect pre-ordered goods to their own addresses – if you order something on eBay, they can get the goods redirected to their addresses, or they can even order goods using your information.

- They can use it to cover their expenses (including medical, travel, and leisure) from your liabilities.

- Attackers can also use other benefits such as redeeming your airline miles to obtain duty-free benefits or cash.

- Attackers can get a mobile SIM card in your name to launch other attacks. When authorities trace the phone number, you will get caught instead of the real attacker.

- They can use your information to open utility accounts with electricity, gas, and communication companies, and then they will make you liable for debts for utilities that you have never used.

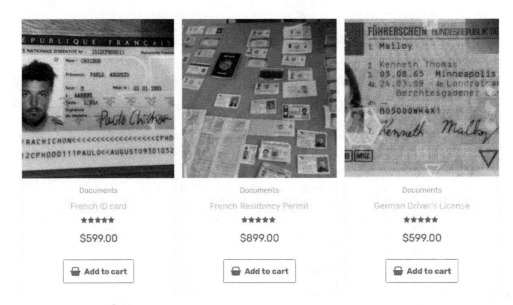

Figure 2.2 – Stolen documents available to buy on the dark web

As well as direct attacks, your information can be used by the attackers for indirect attacks that may create more complex and unpleasant situations. Some examples of indirect attacks include the following:

- Using stolen information, attackers can obtain mobile SIMs that can be used to call other people for different purposes including social engineering attacks and threatening people.

- They can create social media profiles impersonating you and be part of groups in which you have confidential communication to gather information.

- From the social media profiles impersonating you, they can send friend requests to your contacts and then damage your image or lure your contacts to access malicious websites, as your friends may trust them thinking they are you.

- They can launch spear phishing attacks impersonating you. Spear phishing attacks are a type of phishing attack targeting a specific individual or organization rather than everyone.

Some attacker groups may even sell your personal and sensitive information on the dark web. They sell stolen bank accounts, credit cards, and PayPal accounts on the dark web marketplaces.

The following screenshot shows how they sell stolen credit cards and other accounts for low prices on the dark web. There are many marketplaces on the dark web that sell stolen cards, bank accounts, and other financial service accounts for low prices:

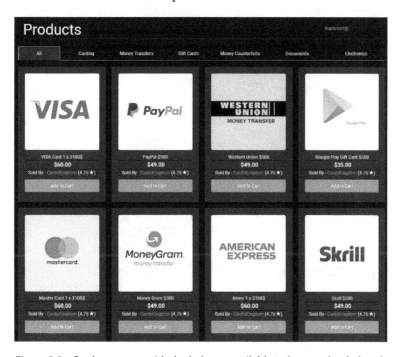

Figure 2.3 – Bank accounts with the balance available to buy on the dark web

Attackers price the stolen data in different price ranges. Social security numbers are treated as sensitive information in the US and are sold at different prices on the dark web.

In the past, there have been massive cyberattacks where attackers have stolen a massive amount of PII data. During a recent attack on Facebook, attackers stole over 500 million Facebook users' data. The most preferred way to use stolen PII data by attackers is by selling it on the dark web. The data sold on the dark web can be used by other groups for fraudulent transactions or committing identity theft.

Individual social security numbers, full names, and dates of birth are usually sold for between $60 – $80 on the dark web and among other hacker communities. The following figure shows a screenshot of things available on the dark web, including driver's licenses and residency permits:

Figure 2.4 – Stolen documents available to buy on the dark web

The preceding figure shows a range of identity documents including residency permits, driver's licenses, and other types of certificates, available to be bought on the dark web.

Technical, procedural, and physical access

We looked at different methods that attackers employ to access personal and sensitive data. We also discussed how attackers can use stolen data for their advantage. It's important to focus on how individuals, governments, and organizations can take countermeasures to protect sensitive and personal data leakage. As a fundamental fact, any security system should protect information, assets, and people physically; physical protection is the foundation of any security system. If physical security is compromised, all other layers we discussed can be compromised quite easily.

When physical security is established, technical security mechanisms can be employed to enforce access control, information classification, surveillance, and monitoring. Procedural protection includes security control, certifications, and badges. Let's look at this in more detail to understand how technical, procedural, and physical access support prevents attackers from assessing sensitive data:

Access Controls	Physical	Fences, gates, and locks
	Technical	Access control systems, VPN, encryption, firewall, IPS, MFA, and antivirus software
	Procedural	Recruitment and termination policies, separation of duties, and data classification

Table 2.1 – Physical, technical, and procedural access controls

Technical controls

When it comes to security, there are diverse types of controls that will be used to control access and usage of data. Technical controls use technological mechanisms to control access to resources and data. You will need to use different technical controls to control access depending on the operating system, resources you want to control access, type of the resource, and the protocol that needs to be used. Many distinct types of technical controls can be used.

Access controls

Access controls are used to authorize identity to access the recourse or object based on the required task. When designing access controls, we need to ensure that the principle of least privilege is ensured all the time, not most privilege access. The principle of least privilege states that a person should be given the minimum privileges required to complete the given task. If the person needs read-only access, they should only be given read-only access to the object. There are mainly three types of access controls in computer-based technology. These three main types of access control systems are **discretionary access control (DAC)**, **mandatory access control (MAC)**, and **role-based access control (RBAC)**.

Discretionary access control

In DAC, subjects were assigned access rights to access objects by the rules. This is usually implemented using access control lists. **New Technology File System (NTFS)** files, a system in Microsoft and Linux, and file and folder security implementation are examples of DAC.

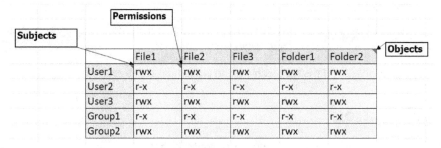

	File1	File2	File3	Folder1	Folder2
User1	rwx	rwx	rwx	rwx	rwx
User2	r-x	r-x	r-x	r-x	r-x
User3	rwx	rwx	rwx	rwx	rwx
Group1	r-x	r-x	r-x	r-x	r-x
Group2	rwx	rwx	rwx	rwx	rwx

Figure 2.5 – DAC permissions table

This is a DAC control table, which will be referred to by the operating system kernel before providing access to any subject to any object. When it comes to implementing this, we will be using permission systems given by the operating system. If you take a Windows 10 system as an example, if you go to the properties of any file or folder, you can configure DAC-based access control. Under the **Security** tab, you can see the list of users including built-in and inherited users that have access to the given object, and the permission given to the users:

Figure 2.6 – NTFS access model

If you click on the **Edit** button, you can provide more specific, least privilege-based access to this object, as follows:

Figure 2.7 – NTFS access model

As per the given screenshot, we can provide more specific, least privilege-based access to the subjects using the DAC model. This can be useful to prevent information leakage if properly configured. Even Linux systems use the same type of DAC technical controls to provide access to the objects:

```
-rw-r--r--  1 root root   241480 Feb 20  2021 wifidemo-01.cap
-rw-r--r--  1 root root      879 Feb 20  2021 wifidemo-01.csv
-rw-r--r--  1 root root      597 Feb 20  2021 wifidemo-01.kismet.csv
-rw-r--r--  1 root root     6753 Feb 20  2021 wifidemo-01.kismet.netxml
drwxr-xr-x  2 root root     4096 Jan 13  2015 .wireshark
-rw-r--r--  1 root root      178 Nov 25  2020 .xml
-rw-------  1 root root     3149 Dec 15 19:49 .xsession-errors
```

Figure 2.8 – Linux file access model

Linux systems use 10 bits to denote permissions for the users and groups to access the object. The first bit denotes the object as a directory (denoted by d) of a file (denoted by -), then the next nine bits are allocated to denote user access, group access, and rest of the world access.

Mandatory access control

From all the technical access controls, MAC is treated as strict access control and is typically used for critical systems and government access control systems. MAC uses the hierarchical method of controlling access to files, folders, or resources. Access to the resources is defined by the system administrators' defined settings. When the access controls are configured using MAC, users will not be able to make any changes, as everything is preconfigured by the system administrators and hierarchically enforced to the below layers.

Role-based access control

Under the implementation of RBAC, other than focusing on the individual subject object-based access discussed in DAC, system administrators assign privileges to the roles. These roles can be predefined or can be created by the system administrators. RBAC is also referred to as **non-discretionary access control**. This gives users access to resources that are required to perform their tasks. One user can be part of one or more roles defined by the system administrator and can be used as and when required to access resources to perform their tasks.

Cloud systems such as Microsoft 365 and Azure mostly use RBAC. Each resource has a predefined set of roles and administrators can assign users based on the required task that the user needs to perform. Administrative roles and resource-based roles can be configured when implementing RBAC:

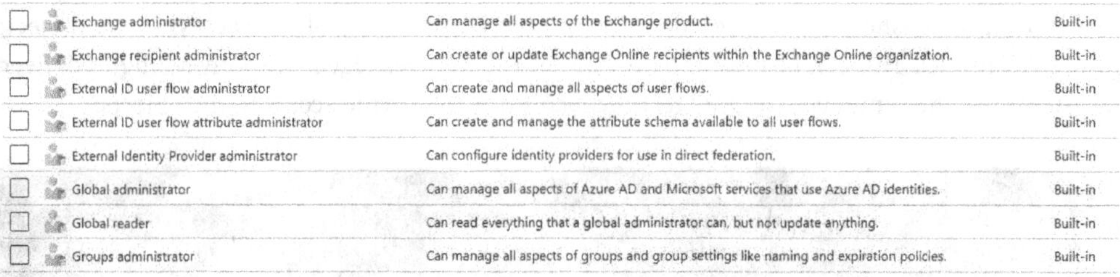

Exchange administrator	Can manage all aspects of the Exchange product.	Built-in
Exchange recipient administrator	Can create or update Exchange Online recipients within the Exchange Online organization.	Built-in
External ID user flow administrator	Can create and manage all aspects of user flows.	Built-in
External ID user flow attribute administrator	Can create and manage the attribute schema available to all user flows.	Built-in
External Identity Provider administrator	Can configure identity providers for use in direct federation.	Built-in
Global administrator	Can manage all aspects of Azure AD and Microsoft services that use Azure AD identities.	Built-in
Global reader	Can read everything that a global administrator can, but not update anything.	Built-in
Groups administrator	Can manage all aspects of groups and group settings like naming and expiration policies.	Built-in

Figure 2.9 – Pre-created roles in Azure RBAC

Simply by assigning users to required roles, users will obtain the required set of permissions to perform the task.

The preceding figure shows the administrative roles, and the following figure shows resource-specific roles that can be used to define the level of access based on predefined roles:

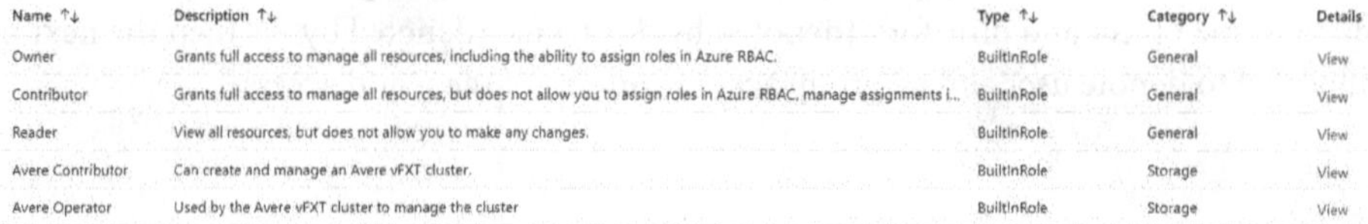

Name ↑↓	Description ↑↓	Type ↑↓	Category ↑↓	Details
Owner	Grants full access to manage all resources, including the ability to assign roles in Azure RBAC.	BuiltInRole	General	View
Contributor	Grants full access to manage all resources, but does not allow you to assign roles in Azure RBAC, manage assignments i...	BuiltInRole	General	View
Reader	View all resources, but does not allow you to make any changes.	BuiltInRole	General	View
Avere Contributor	Can create and manage an Avere vFXT cluster.	BuiltInRole	Storage	View
Avere Operator	Used by the Avere vFXT cluster to manage the cluster	BuiltInRole	Storage	View

Figure 2.10 – Pre-created roles in Azure resources

When implementing technical controls, there are many security solutions, including the following:

- When connecting to branch officers using unregulated connections such as the internet, we can use **virtual private networks** (**VPNs**). These encrypt end-to-end traffic and prevent attackers from eavesdropping on the communication.

- Implementing **multi-factor authentication** (**MFA**). MFA is a mechanism that provides an additional security layer during the authentication based on different factors including something you know, something you have, and something you are.

- Enabling a security policy with a login attempt threshold that blocks the account when a user attempts to log in more times than the threshold.

- Using conditional access policies to restrict access by different conditions including locations, devices, applications, and so on.

- Encrypting data at rest using built-in encryption tools such as BitLocker, and third-party encryption tools such as VeraCrypt to encrypt stored data.

- Ensuring devices are patched and updated with the latest security updates.

- Antivirus and antimalware solutions are properly updated and functioning.

- Device access and usage are properly logged – central logging is preferred.

- If the company allows users to use their own devices – **bring your own device** (**BYOD**) – the company must implement **mobile device management/mobile application management** (**MDM/MAM**) solutions.

- Implementing application-level control. Attackers can use application vulnerabilities and malicious applications to compromise systems, so it's recommended to use application control mechanisms such as AppLocker.

VeraCrypt is an open source data encryption tool that uses strong AES256 encryption that can create encrypted volumes and assign drive letters directly to mounted encrypted volumes. Once you encrypt the data, attackers will not be able to access the encrypted data without the key. VeraCrypt can be used to create encrypted volumes or encrypt existing volumes including system volume. Also, VeraCrypt supports a range of operating systems including Windows, macOS, Ubuntu, Debian, CentOS, FreeBSD, and Raspberry Pi.

Figure 2.11 – VeraCrypt can be used to create encrypted volumes

BitLocker provides built-in encryption for Windows devices but is only supported for specific editions, including the Ultimate and Enterprise editions of Windows Vista and Windows 7, the Pro and Enterprise editions of Windows 8 and 8.1, and the Pro, Enterprise, and Education editions of Windows 10. However, it doesn't support the Windows 10 Home edition:

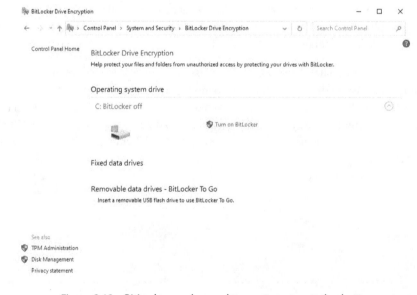

Figure 2.12 – BitLocker can be used to create encrypted volumes

AppLocker is an application whitelisting technology available for Windows devices, introduced with Windows 7 operating systems. AppLocker can be used to restrict applications for the users based on the publisher of the application, application path, and integrity checksum (hash). AppLocker can be configured centrally and enforced using Group Policy:

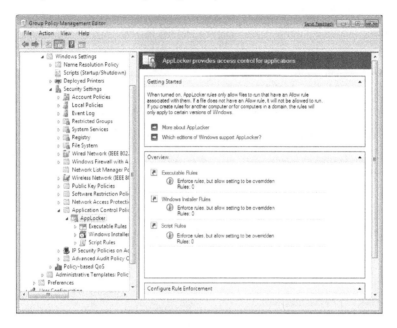

Figure 2.13 – AppLocker can be used to restrict applications

Procedural controls

When you implement technical controls your security posture will be improved, but if users are bypassing technical controls or senior management is not adhering to them, their effectiveness is drastically reduced. This is where procedural controls are crucial.

Procedural controls define what the employees' responsibilities are and how they should behave with the systems. This improves overall security posture and helps with incident prevention. Procedural controls make technical controls more effective in the following ways:

- Enforcing password policies for the organization, including the following:

  - More than eight characters for the password.

  - At least one capital letter.

  - At least one simple letter.

  - A special character.

- Banning common passwords even if they comply with the password policy (for example, Qwerty@123).
- Passwords must be changed after 35 days.
- The last 10 passwords cannot be used again.
- After three incorrect attempts, the account will get locked.
- Prohibiting password sharing.
- Every user must assign a separate account.

- User awareness training to train users on procedures and security
- Implementing disaster recovery and backup procedures
- Recruitment and employee termination procedures
- Communication procedures for sensitive information sharing
- Logging and auditing – this can help during incident handling and forensic investigatory processes
- BYOD and **choose your own device (CYOD)** procedures if the organization accepts these

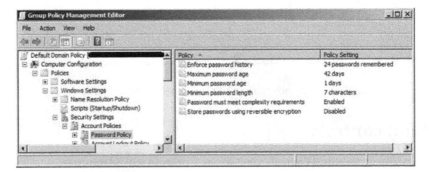

Figure 2.14 – Implementing password policies

Physical controls

Often, the importance of physical control is underestimated by individuals and organizations. But the reality is attackers can attempt to penetrate your infrastructure physically so then they will be able to bypass many other controllers. This must be prevented at any cost as, if attackers get to bypass physical controls, they will have a better chance to compromise other controllers as well. Physical controls include the following:

- Securing devices and physical access controls
- Closed-circuit surveillance cameras

- Motion or thermal alarm systems

- Security guards

- Picture IDs

- Locked and dead-bolted steel doors

- Biometrics (including fingerprint, voice, face, iris, handwriting, and other automated methods used to recognize individuals)

If someone compromises your physical security, you can install different kinds of hacker tools quite easily and get to access your infrastructure.

LAN TURTLE

Sold Out

NOTIFY ME WHEN THIS PRODUCT IS AVAILABLE

Enter your email address...

SEND

DROP A LAN TURTLE. GET A SHELL.

The LAN Turtle is a covert Systems Administration and Penetration Testing tool providing stealth remote access, network intelligence gathering, and man-in-the-middle surveillance capabilities through a simple graphic shell.

Housed within a generic "USB Ethernet Adapter" case, the LAN Turtle's covert appearance allows it to blend into many IT environments.

Figure 2.15 – LAN turtle to get a remote shell

A LAN Turtle is a tool that attackers can use to get access to the shell when they connect this to the network devices. If an attacker physically compromises the infrastructure, they can install this tool in a few seconds.

AirDrive Forensic Keylogger

The **AirDrive Forensic Keylogger** is an innovative ultra-small USB hardware keylogger, only **0.4" (10 mm)** in length. It can be accessed with any Wi-Fi device such as a computer, laptop, tablet, or smartphone. It is the smallest hardware keylogger available on the market, making it a professional surveillance and security tool. The Pro version offers **time-stamping**, **E-mail reporting** and **data streaming**.

$67^{99} or €57^{99}

More info

Figure 2.16 – AirDrive Keylogger

Once the attacker installs this device in a targeted device, they can retrieve anything you type on the keyboard from a close location over Wi-Fi.

AirDrive Forensic Keylogger Cable / Module

The **AirDrive Forensic Keylogger** is a series of specialized hardware keyloggers with Wi-Fi access, aiming at minimizing the risk of exposure. They diverge from the classic USB adapter shape, making them nearly impossible to locate. Available as a **USB extension cable** and **keyboard-embeddable module** only 0.5" (12 mm) in length.

67^{99} or €57^{99}

More info

Figure 2.17 – AirDrive Forensic Keylogger

AirDrive Forensic Keylogger provides keylogger functionality without exposing the connection; this is a USB extension cable shape and is impossible to locate. This is an ultra-compact keylogger hidden in a USB extension cable. This has an ultra-small USB keylogger module that can perform the work of a hardware keylogger.

Creating vulnerabilities to compromise systems

We discussed how attackers compromise systems and steal personal data. This can be done in multiple ways; launching an attack and compromising the system is one of the ways that attackers steal your data. To launch an attack, there must be three components to be fulfilled.

They are as follows:

- Vulnerable system
- Exploit for the recovered vulnerability
- The motivation of the attacker

Vulnerable system

The definition of a vulnerable system is a system with existing weaknesses. This is not only an operating system. There can be different layers of the system that can be vulnerable. It can be an operating system or installed applications, in which the procedures may allow for creating vulnerability, and the protocols used can be vulnerable. This can be recovered by vulnerability assessment. During vulnerability assessment, testers will try to understand whether the targeted system has any known vulnerability. Testers can use automated tools such as Nesses Pro, Core Impact, Open VAS, and SecPoint to scan the targeted system to recover vulnerabilities. There are plenty of known vulnerability databases. The **National Vulnerability Database (NVD)** (`http://nvd.nist.gov`) is a known vulnerability database:

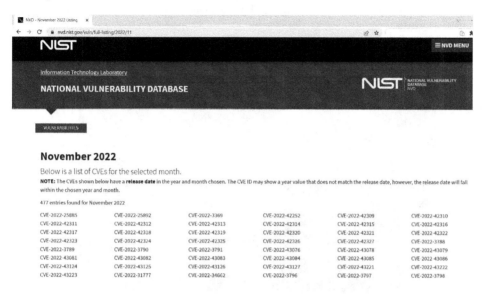

Figure 2.18 – National Vulnerability Database

From the security point of view, we need to maintain all our systems free of vulnerabilities. If vulnerabilities cannot be found, it's much harder for an attacker to compromise systems. Then, the attacker must find a way to create vulnerabilities.

Exploit for the recovered vulnerability

Once the vulnerability is recovered, then the attacker must find an exploit to compromise the vulnerable system. An exploit is a defined way of compromising the system. There are databases and frameworks that attackers can use to find readily available exploits, or attackers must develop their own exploits.

If you access the Exploit Database (`https://www.exploit-db.com`), you can find many exploits developed for known vulnerabilities:

Figure 2.19 – Exploit Database

Exploit Database is a known exploit database that's contributed to by many communities. They are constantly updating the exploit database with developed exploits. Attackers often use these exploits to launch attacks over known vulnerabilities. They can use also some frameworks, such as the Metasploit framework:

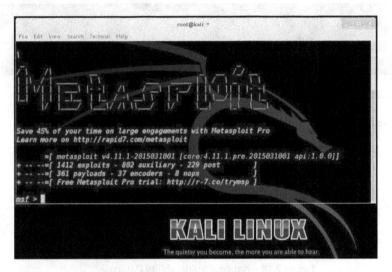

Figure 2.20 – Metasploit framework

The motivation of the attacker

To launch an attack, the motivation of the attacker is important. The reason is that even if the system is vulnerable and exploits are present, if the attacker is not motivated, an attack will not take place. Sometimes when we propose security assessments for companies, they say there is no requirement for testing as they are fully secure. No one can say they are fully secure as there can still be ways attackers can exploit your systems.

Previously, we discussed vulnerabilities of the target systems and a few databases that we can research for vulnerabilities. Still, there can be undiscovered and unpublished vulnerabilities that can lead to compromise of the system. When systems engineers and security teams suggest security solutions or improving existing security systems, sometimes management may deprioritize the requirement stating that they have been using older systems for a long time and have never been attacked. The reason can be that attackers may not be motivated to launch an attack on the infrastructure.

As we discussed, even if the systems are vulnerable, attackers will not try to exploit them if they are not motivated. Simply, attackers would wonder whether the attack was worth it. The main motivation for an attacker is money but sometimes, taking revenge or destroying the company's image can be other types of motivation. Once, there was a guy who compromised an enterprise infrastructure just to see what his girlfriend was doing. These are various kinds of attacker motivations.

Out of the three components of attack, exploits, and motivation of attackers, we have no control. The only component that we have some level of control over is the vulnerabilities of the system. This is the reason we must conduct vulnerability assessments to recover existing vulnerabilities of the system.

Creating vulnerabilities

If the attackers have higher motivation to compromise a system and if they cannot recover any vulnerability, they will not be able to exploit it. Then, the only option available for attackers is to create a vulnerability in the system. Attackers use different ways to create vulnerabilities in your system and compromise it. These are some of the ways attackers create vulnerabilities:

- **Trojans** – Trojans, or Trojan horses, are types of malware that pretend to be legitimate software and are often infected through email attachments or malicious websites. Once infected, a Trojan provides attackers with access as a backdoor to the system.

- **Phishing** – Phishing is a type of social engineering technique where the target user or organization is misled via email or chat services. Once the user clicks a link in the email or chat, they will be redirected to a specially crafted website that looks legitimate, and the attacker will be able to steal sensitive data, typically credentials.

- **Malicious insiders** – If the infrastructure or system is not vulnerable, attackers might use malicious insiders or disgruntled employees to create vulnerabilities in the system.

There are many tools used by attackers to create vulnerabilities. Attackers use Trojans in many operating systems including Windows, Linux, iOS, and Android. Once the system is infected, attackers have access to the system. As an example, AndroRAT is a Trojan that can be bundled into APK files or used to create binder APK and installed on the Android device, which gives total access to the attacker. AndroRAT Binder will be used to create the apk installer, and then the APK file must be installed on the Android device:

Figure 2.21 – AndroRAT binder

When the APK is installed on the Android device, it will create a reverse connection to the attacker's system that has AndroRat running. Once the connection is established through the configured port, it will be shown like the following screenshot to the attacker:

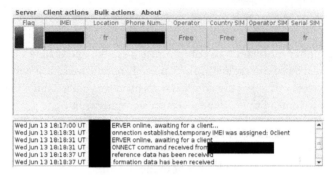

Figure 2.22 – AndroRAT control center window

When the connection is established, the attacker can take complete control of the device by double-clicking on the selected device if there are multiple connections to the AndroRAT control center:

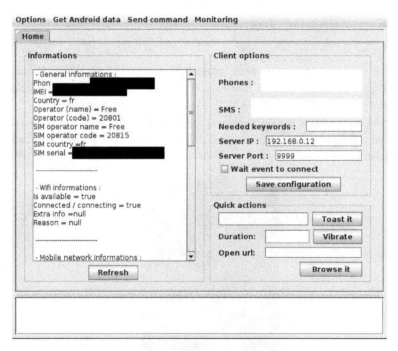

Figure 2.23 – AndroRAT connection window

Once the attacker is connected to the device, there are many functions available for the attacker, including getting access to the Android device data, generating calls, sending SMS, vibrating the phone, and toasting the phone.

As well as Trojans, commercial apps are available to provide silent access to mobile devices. If you receive a mobile device as a gift, especially if the device is not a bubble wrap (bubble wrap is a term used to denote the device is unopened or sealed), there is a chance that the device has spyware or antitheft apps installed that gives access to your phone to a third person. At the very least, we need to factory-reset such devices before starting to use the device.

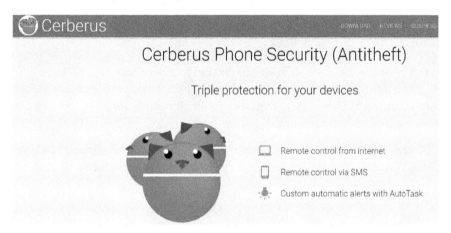

Figure 2.24 – Cerberus Phone Security

Cerberus is capable of a rich set of features that any attacker would expect of a good **remote access tool** (**RAT**), including locating the device, backing up data, data exfiltration, locking the device, changing the PIN, wiping data, and taking photos. Interestingly, all these capabilities and many other features will be available from any web browser or by sending special SMS text commands.

Increase the attack surface using sensitive data

Previously, we discussed the main elements of attacks including vulnerabilities. If the vulnerabilities cannot be recovered, attackers could try to create vulnerabilities. From the users' perspective, we need to reduce vulnerabilities as much as possible. The challenge is that there can be vulnerabilities in the system that have not yet been recovered or published. We can only remove vulnerabilities that are known. Unknown or undiscovered vulnerabilities are typically referred to as **zero days**.

Even though we conduct vulnerability assessments, we conduct an assessment based on known vulnerabilities. To battle against zero days, we need to reduce the attacker surface. An attacker surface is a set of entry points or boundaries in the environment that attackers can use to try to enter, create connections, and exfiltrate data from the systems, system components, or infrastructure. To increase security, we need to always try to decrease the attack surface.

An attack surface can comprise the following:

- System endpoints, such as workstations, laptops, and mobile devices used to access your environment or infrastructure

- Servers in the system including domain controllers, file servers, application servers, and database servers

- Routers, firewalls, and switches

- Printers and other network-connected devices including all-in-one printers

- Identity systems, including Active Directory Domain Services or **lightweight directory access protocol (LDAP)**

We need to reduce the attack surface as these endpoints can introduce potential attacks on the system. For example, if you have **remote desktop protocol (RDP)** (which is used to establish remote access to systems) access enabled in your server that has a public IP address assigned, attackers can easily locate these servers and try to compromise them.

If you access Censys (`https://search.censys.io/`) and search for Windows 2003 RDP, you will find a list of Windows 2003 servers with RDP enabled, which increase the attack surface:

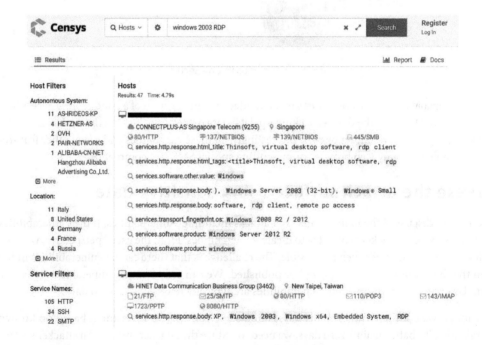

Figure 2.25 – Censys search

Once you try to connect, it will give you a connection prompt. This shows very clearly some configurations increase the attack surface.

Attackers go through five phases to launch an attack, as follows:

1. Information gathering or reconnaissance – gathering information as much as possible will help an attacker launch an attack on the target more accurately

2. Enumeration and scanning – this phase is to collect more specific information about the services and the ports on the target, which allows the attacker to recover vulnerabilities and map the target with exploitations

3. Gaining access – this is the real exploitation phase

4. Maintaining access – if the attacker needs to connect to the target another time, they do not have to go through the time-consuming information gathering and scanning phases if the attacker plants a backdoor on the compromised system

5. Covering tracks – the attacker removes all traceable entries on the compromised system

From these phases, the first two phases are typically time-consuming, as the more information is collected, the more possibilities the attacker has to compromise the system.

Overexposure can reduce the time that the attacker needs to spend for the first two phases. If the attacker collects information in the ways that we discussed earlier, this reduces the time for attacks. For example, some users overexpose system information on social media such as LinkedIn, which can reduce the attacker's time.

Sometimes, people publish selfies taken during a candlelit dinner, and exposing their official identities or credit cards on the table can be used by attackers to launch attacks, as this sensitive information increases the attack surface.

The same thing can happen with the information collected by applications, browsers, and social engineering techniques that we discussed earlier. Attackers can find a list of people working for specific organizations using a tool such as **TheHarvester**:

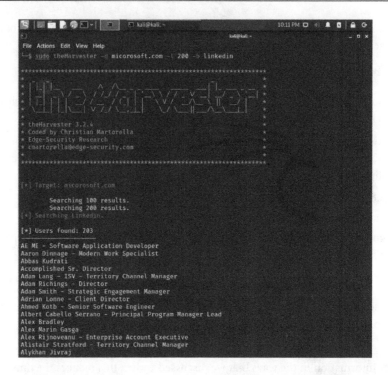

Figure 2.26 – TheHarvester search for employees working for Microsoft posted on LinkedIn

The preceding figure illustrates how easily attackers can find out the people working for specific organizations just by executing a simple command.

Creating entry points and bots

Attackers use different types of entry points to enter systems and infrastructures. As we discussed, the attack surface can be used by the attackers as entry points to the systems through different methods and attacks, including the following:

- **Compromised credentials** – Typically, your username and password are treated as the primary method of authentication. You can enable MFA to increase security. If the credentials are exposed to attackers, they can use the credentials to access the systems, even if these credentials do not provide administrative access. Compromised user accounts can be critical as they represent legitimate user behavior for most security systems. As well as users, there can be service accounts with the *Password never expire* option enabled, and local administrative accounts with the same password on all the systems that never change the password. These can be easy targets.

- **Common and weak credentials** – Even though we configure complex password policies, some users can use common passwords as they fulfill the basic requirement to become a complex password. Examples are `Qwerty@123`, `Admin@123`, `Test@123`, and `Asd@123`.

- **Missing and poor encryption** – Encryption converts plain text to cipher text, which makes it harder to understand by humans. Typically, all passwords are stored as a hash. Even if someone gets to access the hash, they will not be able to get the real password. We discussed how encryption can be used to store data, but also, encryption can be used to establish secure data in transit. If the encryption is not present, passwords sent as plain text can be easily compromised by attackers by launching man-in-the-middle attacks:

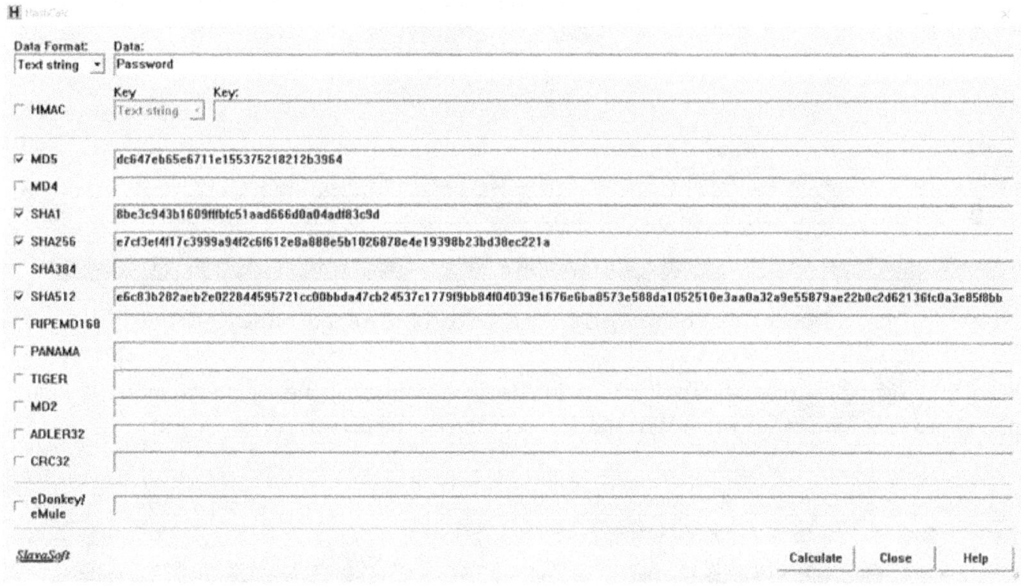

Figure 2.27 – Hashcalc generates hash codes for given plain text input using multiple algorithms

Password files will store encrypted passwords without storing human-readable plain text passwords. If the passwords are not encrypted, attackers can see the communication in plain text. As an example, when you communicate using `http` or `ftp` kinds of protocols, they don't encrypt the communication, and attackers can easily intercept the passwords in plain text.

As an example, I'm using a Joomla CMS-hosted web application that is hosted without a digital certificate, which is a provided web service through `http`:

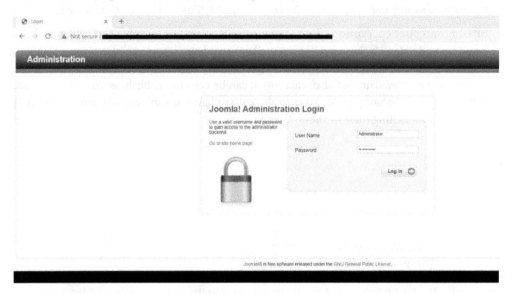

Figure 2.28 – Joomla web application backend without https

Whenever a user tries to access this web application backend, credentials are sent to the server as plain text. Let's see what we can see when **Wireshark** is running on the network:

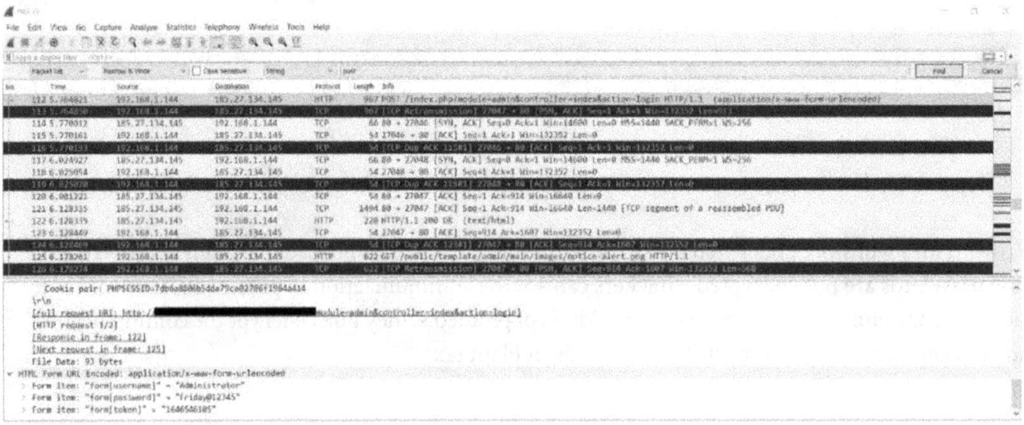

Figure 2.29 – Wireshark captured the web application credentials

The list is as follows:

- **Malware** – Attackers can use malicious code to compromise the systems; malware can take over systems, exploit vulnerabilities, and even create backdoors once infected.

- **Password attacks** – Attackers use several types of password attacks including dictionary attacks, brute force attacks, keylogger attacks, credential stuffing attacks, and password spraying attacks. Implementing MFA, conditional access, and hardened password policies can prevent these types of attacks.

- **Castle security implementation** – Castle security implementation is something like the security arrangements in the castles in the olden days. In castle security, total strength is given to the castle perimeter to protect the castle. Rigorous checking is carried out at the gate and any entity that goes through the gate is typically treated as trustworthy. Even older system implementations have the highest attention given to protecting the perimeter. Most of the appliances were implemented at the perimeter such as firewalls, **intrusion detection system/intrusion prevention system (IDS/IPS)**, and **security information and event management (SIEM)** solutions. But this will not protect against attacks from insiders. Currently, the accepted implementation is known as **zero trust**. When implementing zero trust, no assumptions are made based on the location or access requests; all requests will be verified before providing access.

- **Default configurations and passwords** – Many devices implemented today come with default passwords. Often, administrators don't change them. This leads attackers to locate the devices connected to the internet quite easily and access them using default passwords. If you visit `https://datarecovery.com/rd/default-passwords/` and `https://www.cirt.net/passwords`, you can find a long list of default passwords to many systems.

If you search for any appliance on `https://www.shodan.io/`, you can find real devices connected to the internet. Let's search for Hikvision:

Figure 2.30 – Shodan shows internet-connected Hikvision devices

When you click on any of the devices, it will take the admin login to the device. Most of the devices can be accessed with the default passwords that we discussed earlier. Once attackers gain access to the device, they can monitor, make changes to the device, and even forward ports to existing devices that are connected to internal networks.

Zombies and botnets

Bots is simply the short name for **robots**, which refer to automated programs or scripts. Once infected, your system will become a compromised host that can be controlled by the attacker, known as a **zombie**. Infected machines can be remotely controlled by attackers. Even though it's a tiny program or script, by combining thousands of them, attackers can often bring down large systems using **distributed denial-of-service** (**DDoS**) attacks. A large number of bots is referred to as a **botnet**. Often, these scripts are infected in your browsers, programs, and even devices.

Summary

In this chapter, we discussed why attackers are interested in your privacy and the benefits for attackers from an attacker's point of view.

We discussed the ways that attackers use stolen data for their benefit, and how we can protect our data using technical, procedural, and physical access controls. We discussed why recovering and mitigating vulnerabilities are critical as the vulnerabilities can be used to exploit systems. When vulnerabilities are not present, attackers can create some. We looked at how the reduction of the attack surface can impact potential attacks. Overexposure increases the attack surface. Attackers use different entry points to access systems using various technologies. Bots are small programs or scripts that can infect your system and compromise the system.

In the next chapter, we will learn how attackers reveal the privacy of individuals and companies.

Part 2: Methods and Artifacts That Attackers and Competitors Can Collect from You

This part explains how attackers and competitors can collect information from you and what type of tools they would use.

This part comprises the following chapters:

- *Chapter 3, Ways That Attackers Reveal the Privacy of Individuals and Companies*
- *Chapter 4, Techniques that Attackers Use to Compromise Privacy*
- *Chapter 5, Tools and Techniques That Attackers Use*
- *Chapter 6, Artifacts that Attackers Can Collect from You*

3

Ways That Attackers Reveal the Privacy of Individuals and Companies

In todays digitalized world, most of the information are exposed without individuals and companies' knowledge. This has become a new landscape which presents new risks and threats to individuals and companies that must be remediate. Otherwise, exposed information can be increasing the attack surface and attackers can take the benefit. Let's deep dive into the ways that attackers obtain this information by over exposure. As explained in earlier chapters information has become power and currency.

This chapter explains how attackers revealing privacy of individuals and companies. In this chapter you will be able to understand:

- Cyber exposure index
- Everything exposed in cyberspace matters
- We don't need to enter our real information always
- Facts we shouldn't be disclosing
- Social media matters

Cyber exposure index

Cyber exposure index provides indicative value of how much information a company has exposed based on sensitive disclosure, exposed credentials and hacker group targeting. Cyber exposure index is defined based on the data collected from publicly available data sources in the dark web, on the deep web, and in data breachers. Cyber exposure index shows how the company is ranked based on the following variables:

- **Sensitive disclosure** – Typically sensitive disclosure regulated by laws and policies and never be stored on the devices or shared without proper authorization as sensitive information consist of confidential data, trade secrets, business plans and other valuable information. When the sensitive data disclosed, disclosed information can be used for identity theft and other attacks.

- **Exposed credentials** – Usually exposed credentials can be usernames, passwords, tokens, or any other forms of identities that provide access to critical systems. This can be critical as exposed credentials are the mostly used way of compromising systems and provide hackers with access to critical systems. Attackers can collect credentials from compromised systems, information leaks and other attacks including social engineering.

- **Hacker group targeting** – Hacker groups are organized gangs or communities that can actively share stollen credentials and other information among them and launch collective attacks. Mostly these attacks can be launched from different geography's that makes harder to traceback the real locations. Also, they can collectively launch distributed denial-of-service attacks to bringdown the cooperate and critical systems which prevents legitimate users access to the systems.

there are various ways that individuals and organizations expose their sensitive information to public sources unintentionally. This can be used by attackers for their benefits by launching direct attacks or use them for indirect attacks. Unintentional information exposure can happen through WHOIS records, Email addresses, Contact numbers and various other ways.

WHOIS records

Organizations and individuals often overexpose sensitive information intentionally or unintentionally. WHOIS records are publicly available information on internet entities. A company or individual can obtain a domain, and domain registration related information publicly available. By running a simple query or using an online tool, attackers can easily collect this information and use them for malicious activities. If you use online service www.who.is and search for preferred domain, it will return you the information including registrant contact information, administrative contact information and technical contact information. This information may be personnel contact information:

Figure 3.1: whois search for Microsoft.com domain

As you can see WHOIS database query shows the information related to the internet resource including contact information and organizational information.

You can prevent disclosing personnel information on WHOIS by enabling WHOIS privacy controls. If the domain service provider provides privacy controllers, we can prevent disclosing personally identifiable information. You should avoid entering false information on WHOIS data as it can revoke ownership for the domain. Best option is to enforce privacy controls to WHOIS information for the domain. Apart from that there are many other concerns that we should prevent disclosing our information:

- **Increasing spam** – if the email address is visible on the WHOIS record, it increases spams as spammers can easily collect authentic email addresses from WHOIS records.

- **Entity mapping** – when your information disclosed on the WHOIS database, attackers can map you with other information sources including social media and collect more information about you and map them to other attacks.

- **Identity theft** - when the authentic information is publicly available attackers can easily use them for Identity theft related attacks.

- **Disclosure of the ownership** – WHOIS records discloses the ownership of the site. Attackers can launch spier phishing attacks and more targeted attacks when the ownership is disclosed.

Considering these factors enabling privacy control and protecting from public disclose will prevent many attacks. Enabling privacy controllers is not expensive. Domain privacy controllers service fee is not very expensive. Most of the service providers provide these services for less than 20$ per year.

Email addresses

Email is one of the most used piece of personnel information for many requirements. Specially creating social media accounts, registering for government services, registering for banking services and many more. When using same email address to multiple services, it introduces a risk. One of the known risks is compromising email address can provides access to all the services. When registering to blogs, newsletters, and many other types of subscriptions email address increases the visibility to many services. As a result, user might start receiving spam mails, hoax mails, and emails with malicious attachments and targeted to phishing campaigns.

One of the countermeasures is creating email alias. Email alias is creating different email address for same mailbox. When a mail is sent to email alias, it will be received by the user's primary mailbox. Many email service providers support mail aliases. Some providers only allow limited number of aliases to be created. It's easy to create, maintain and work as a temporary email address. This reduces the risk of introducing same email address for all the services. Specially for newsletter subscriptions and related temporary services we can use email aliases and later we can remove when not required.

The following are some email alias service providers that we can use:

- **AnonAddy** – provides anonymous email forwarding.

- **Firefox Relay** – provides limited number of aliases.

- **SimpleLogin** – privacy focused opensource project provides email alias.

These services provide alias services where you can protect your inbox from spams, and you can be anonymous in the cyberspace. The following screenshot is a reference to SimpleLogin service:

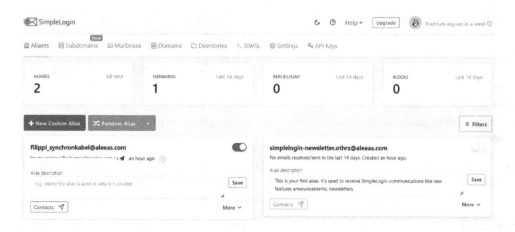

Figure 3.2 – Create custom and random alias using SimpleLogin

As the preceding example shows, SimpleLogin can create custom and random alias to your personnel or official email address to prevent disclosing your real address. These email aliases can be created, enabled, or disabled. This also shows the number of emails received or sent to monito email related activities.

SimpleLogin is opensource project which provides unlimited email aliases. It also offers iOS and Android apps for easy management. SimpleLogin support custom domains and **Pretty Good Privacy (PGP)** encryption. Additionally, SimpleLogin also provides browser extensions for Chrome, Firefox, and Safari.

If you are using corporate solutions like Microsoft 365 or Google workspace, you can manage this on admin portals and create any number of aliases you need and manage accordingly.

Mobile numbers

After protecting email address from disclosure, another very important personnel record is your phone number. There are many ways that attackers can recover your phone number. The following are some of the common ways that attackers can discover your phone number:

- WHOIS records.
- Social media accounts attackers can disclose phone numbers (Example, LinkedIn).
- From your resumes submitted to job portals.
- Your phone bill thrown out to the bin.
- From your email signature.
- Sending you spam mails and when you reply with your email signature.

It is important to remember that most of your security capabilities relying on your phone number. Even if you have strong passwords and **Two-Factor Authentication** (**2FA**) enabled, you are relying on your phone number. Imagine someone who has access to your phone number can bypass anything. For example, anyone who steals your phone number can easily reset your google password just by sending google code. The same technique applies to social media accounts, bank accounts and many others. Basically, the attacker who steals your phone number can virtually become you. As most of the services including government services identify you by your phone number, anyone who has access to your number can impersonate you.

Interestingly we try to protect our bank account numbers, social security numbers and credit card numbers but are we protecting our phone number in the same way? There are ways that attackers can steal your phone number and receive all the calls and text messages sent to your phone number.

Once they get your phone number, they call the service provider impersonating the customer. Most of the telco providers will *port out* the phone number to different SIM card after they verify you with series of questions. Answers to most of the questions can be found using information gathering techniques. In some countries even attackers can *port out* the number to different service provider and a different SIM online. In many countries such SIM cards can be bought from supermarkets. When the *port out* is completed, the new SIM will receive all the calls and text messages sent to the target number.

To mitigate this risk, we need to deploy additional layer of security like 2FA for the phone numbers. The options available are different from country to country and service providers. Some service providers provide different methods like secondary password attached to the number, which is required if you want to port out the number to another service provider.

Another way of protecting your number is using built in feature like *blocking the caller ID* from your phone. Many operating systems including iOS support blocking caller IDs. Once you make a call, recipient will not see your number. But if the recipient number is not available, some of the service providers send call alert with the caller number even when the caller ID is disabled.

Disposable phone numbers can be used to overcome this. Virtual numbers or secondary numbers are another option to make and receive calls securely. Virtual number can be changed whenever you want. Even you can use numbers from different countries if you use virtual numbers. There are many apps available to obtain virtual numbers. We will discuss disposable, virtual, and secondary numbers in detail on upcoming chapters.

Everything exposed in cyberspace matters

Once we are connected to the internet – which is the case at any given time, everything we share, search, download, access is collected, monitored, stored by multiple entities. These entities can be operating systems, applications, browsers, network connections and even the resources that we access. Even though the public go with *I have nothing to hide* attitude, hackers, spies and even governments may be interested in all your activities.

This is the reason, some of the countries have built massive filtering systems to monitor and prevent exposing sensitive information. As an example, China has built the *Great Firewall of China* which is a stronger censorship which prevents access to selected foreign websites and slowdown cross border internet traffic. Using this link, you can check which websites are blocked in China, `https://www.greatfirewallofchina.org/`. This will make accessing censored website more difficult from China using censorship circumvention technologies like VPNs for the blocked websites including Facebook, Twitter, Google and YouTube.

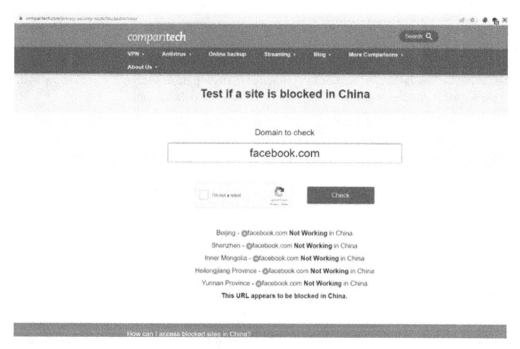

Figure 3.3 – greatwallofchina.org showing the blocked websites in China

Internet activities monitored in different levels including **Internet Service Provider (ISP)** level; all internet activities can be monitored by ISP. Even ISP level attacks can be conducted by the attackers and the services can be hijacked. This is not a common attack but the sites you visited can be tracked by the cookies. Cookies are small piece of text data created by the web server or web application within the web browser and stored in the device you access the website. Typically, cookies are used to provide better internet experience including items suggestions on a sales catalog like eBay. Once you search for a specific item, this information stored in cookie session data, and later the web application can refer to the previous data and suggest you a similar item. Also, cookies are used to target users for tailored advertising.

Exposing an email address

Generally, we think our email address is something that we can share freely with everyone. But we've forgotten our email address is the entry point for most of the services we use to access including social media and banking. Mostly our email address is an identity for many other services. It's not a surprise that attackers try to compromise email addresses. If we compare this act with previously discussed hacker phases, finding the email address of a target can be treated as an information gathering. If an attacker knows your email address, that increases attack surface.

As a countermeasure we can use multiple email addresses and share relevant email address when required. Specially email address connected to your sensitive services must be kept confidential. And keep separate email address to social media, blog posts and other services which reveals your email address to public. This is the reason, attackers interested in finding email addresses posted in internet, there are ways that attackers can find email addresses posted in social media.

As an example, let's say you want to find out people who are working as HR managers with Gmail address published in LinkedIn can be recovered by running simple Google dork that we discussed earlier. We will discuss Google dorks in detail in *Chapter 5, Tools and Techniques That Attackers Use*. But as of now you will see, when executing below google dork will output list of profiles in LinkedIn, working as HR Managers and has gmail.com email addresses on their LinkedIn Profiles. In the same way you can change the criteria and list out profiles with email addresses published based on the job title, country, language, and various other type of attributes published on LinkedIn.

The following statement shows a Google dork crafted for the criteria we discussed:

```
http://www.google.com/search?q=+"HR+Managers" -intitle:"profiles"
-inurl:"dir/+"+site:www.linkedin.com/in/+OR+site:www.linkedin.com/
pub/ "*gmail.com"
```

If the email address is Gmail, and if the email address is published on LinkedIn, attackers can simply list all LinkedIn users within the criteria defined, then attackers can copy all the google findings and use online email extractor services to extract only the email address from the copied text. As an example, attackers can use `https://email-checker.net/extract-email`. Attacker can select all the content or simply pressing (*ctrl + A*) to select all Google search result and paste it on this service and click extract email.

This tool extracts email addresses from the text data. Similarly, there are different types of tools that can extract specific data including contact numbers, locations, first name and last name.

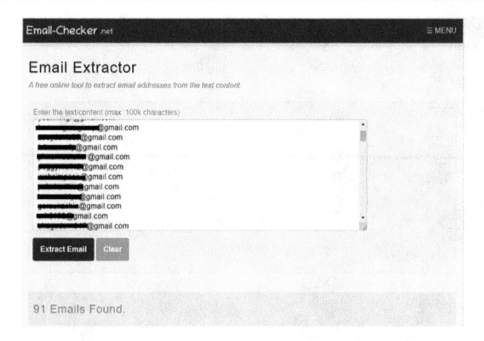

Figure 3.4 – Email extractor extracts only the email addresses from pasted text

This online tool then extracts email addresses in the text, same technique can be used by the attackers to extract email addresses on websites, blogs, and other resources.

Other than commercial and online tools attackers commonly use grep tool in Linux to extract information using regular expressions. Command grep supports regular expressions.

The following command will extract email addresses from a text file call junk_text.txt:

```
grep -o '[[:alnum:]+\.\_\-]*@[[:alnum:]+\.\_\-]*' junk_text.txt
```

[[:alnum:]+\._\-]*@[[:alnum:]+\._\-]* is a regular expression that matcher email addresses.

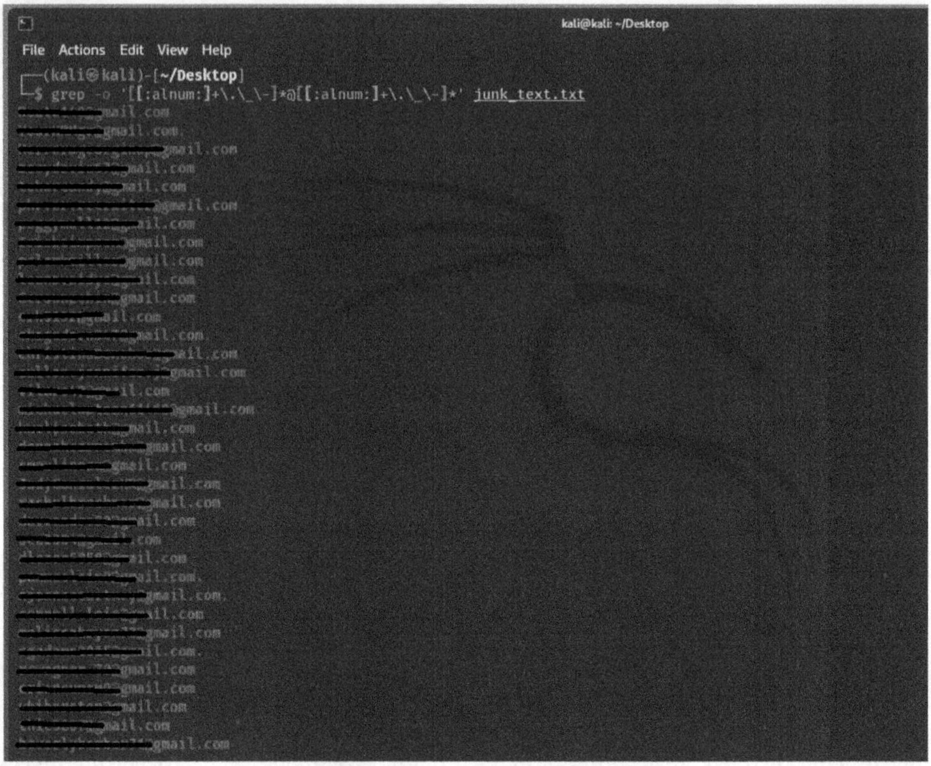

Figure 3.5 – Email addresses extracted

Not only email addresses, but attackers are also interested in your phone numbers or any kind of other contact details as well. Later attackers can use them to launch social media attacks. Some targeted attacks, attackers impersonate service providers and try to change phone numbers using social engineering techniques. Then they can even bypass multifactor authentication **One Time Password (OTP)** if successful.

African emperors' treasure

The famous Nigerian Prince story is one of these techniques to collect authentic information that can be used for attacker benefits. Attackers contact random contacts and send emails claiming the sender is one of the great grandsons of African emperors and looking for someone to transfer his/her inherited treasure and asking information including full name, contact number, email address and bank account information. Sometimes they even attach a picture of them to make the story real. Some users share their information as they believe the story or sometimes, they think that no harm sharing information with the recipient. Users often think if the story is true, I will get the money, even if the story is not true there will be no harm.

Figure 3.6 – Sample for similar attack requesting information to transfer huge fund

But for attackers' point of view information is the most important part. When they get authentic information about full name, contact numbers, email address, postal address, and banking information, they can easily impersonate you to the bank. By impersonating you, they can request bank to change the postal address and request for ATM cards. When the bank sends out ATM cards and PIN numbers to new address, attacker can withdraw cash from your ATM card, or they can purchase online services or buy crypto currency using your card. Usually, attackers change your postal addresses to commonly accessible postal addresses. For example there are apartments with insecure and unprotected post boxes located in the ground floor or share insecure common post boxes. When the attackers found these types of unattended and unprotected postboxes, they can use them to redirect important letters to unattended postboxes and collect them without noticing to anyone. Then attackers changes the postal address of the target to this type of unprotected postboxes, they can collect the postal mails sent to the target user without noticing anyone.

Nigerian prince story and similar attacks are types of information gathering where attackers collect legitimate and authentic sensitive information for their benefit. Attackers use this information to launch direct attacks or to be used for indirect attacks. **Maltego** is a commonly used tool among Penetration testers, Security researchers, forensic investigators, and hackers, for information gathering. We will discuss in detail about Maltego in the *Chapter 5, Tools and Techniques That Attackers Use*.

IP addresses

We use IP addresses to communicate with the devices connected to network. Devices can have different form factors and different operating systems. We use private IP address to communicate with internal devices and we must have public IP address to connect to internet. If you have multiple devices connected to your home network, typically all the devices go through single public IP address assigned to the router. Public IP address is assigned by **ISP**. Your IP is your path to internet. All the communication will go through the public IP address. Public IP address can be assigned dynamically or statically. Static IPs usually assigned for specific services like web server to publish web service. Generally, home connections use dynamic IP addresses. Dynamic IP addresses are changing frequently but static IP addresses will not be changed.

If you open command terminal window and type `ipconfig` or `ipconfig /all` will show your internal IP information.

```
Wireless LAN adapter Wi-Fi:

    Connection-specific DNS Suffix  . :
    Link-local IPv6 Address . . . . . : fe80::254e:f08b:c790:5554%11
    IPv4 Address. . . . . . . . . . . : 192.168.1.144
    Subnet Mask . . . . . . . . . . . : 255.255.255.0
    Default Gateway . . . . . . . . . : 192.168.1.1
```

Figure 3.7 – Internal IP configuration

Attackers are more interested in your public IP than your internal IP as Internal IP address is only useful within the network. You can find your public IP address by just opening your web browser and typing `whatsmyip` on the google search or accessing `https://ip4.me` website.

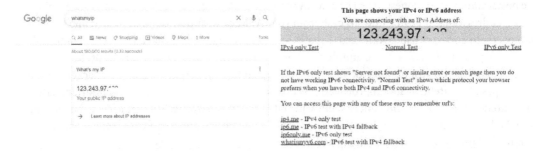

Figure 3.8 – Public IP configuration

Your public IP address is the path to locate you even if it does not expose your personnel data. Often public IP is not treated as **PII** hence companies does not take extra precautions to conceal public IP.

If you are connected to the internet via home router, your public IP address is associated to the WAN link. Often attackers use multiple ways to extract your public IP including:

- **Email communication** – some email providers does not mask your public IP, when you send email, attackers can obtain your public IP by analyzing email header.

- **Online advertisements and accessing websites** – when you access a website or click online advertisement it collects your public IP.

- **Internet sharing and torrents**- when you are part of file sharing or torrents, your peers will see your public IP address.

Your public IP address will not reveal your personnel information, but attackers can obtain location information by IP address as your ISP assigns your IP address. If attacker collects your Public IP address, attacker can:

- Attackers can use your public IP address to ban you from some services including online games and emails.

- Your IP address reveals your location.

- Attackers can use your public IP address to impersonate you.

- Your will be framed for crimes that you never performed.

- Can try to compromise your router and gain access to internal services and information gathering.

When an attacker knows your public IP address, they can scan your IP address and try to find the device that is connected you to internet and open ports and running services of the device. Most of the ISP's configure devices with default passwords, attackers can try to access your route through WAN link. For an example if the attacker enters your public IP into a browser, typically attacker can see the login information of the router.

Figure 3.9 – Login screen of your home router over the public IP

Then attackers can try to obtain access to the router using default passwords assigned to this router model or using password cracking techniques. If the attacker successful, attacker can find information of the connected devices, users, MAC addresses of the devices and even your SSID passwords.

We don't need to enter our real information always

As described earlier, any information shared in the cyberspace matters. We never know what types of **Open-source intelligence (OSINT)**sources have access to them. Anything shared on social web applications, blogs, or even simple email can increase attack surface. Often, we disclose personal information like full name, postal address, email address, hometown, mother's maiden name unnecessarily. On the other hand, by using accurate information we are risking ourselves.

Let's say there is a document that you want to download from a social site, but it demands to register to get you access to the download link. During the registration process it often required your full name, address, contact number and the email address. Your only intention is to download the document, but it is required to register by providing all these information to get the link. Do you think it is necessary to provide accurate information to get download link? Probably, this can be an attacker's trap to collect your information. This can be social engineering attack. Download link can be either fake or real. But it required to provide your information to get the link.

Let's see the ways to overcome this kind of an information leakage. We can typically provide false information on full name, home address and contact number but email address is the challenge as often the link will be sent to your email address. We will take this question in two parts; one is to provide your information. There are many sources that provide test data to be used in this type of situations without providing your real data. `https://dlptest.com/sample-data` is an example for a public data source to provide sample data:

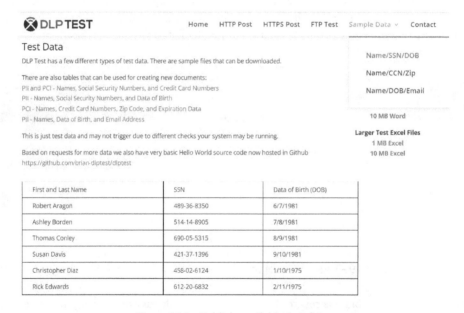

Figure 3.10 – Publicly available test data

This site provides sample data including full name, **Social Security Number** (**SSN**) – unique identifier assigned for US citizens to track income and other government benefits), date of birth and credit card information.

Next challenge is providing your email address as without email address you will not receive download link or activation of your account. One method is to create temporary email address or email address with false information and can be used to this type of temporary registrations. We can use free email address provider to create a temporary email address like Gmail, Yahoo or Live. But these email providers often need another email address or your contact number to create temporary email address for you. Some of the email services like `https://protonmail.com/` and `https://www.mail.com/` can be used in some countries without additional information to create temporary email address. Other option is to use disposable email address.

On the other hand, providing accurate information when not required can increase attack surface. For an example when you provide answer for security question to reset the password for your personnel email address, you are risking your credentials. Some of your personally identifiable data are not confidential for your close circle. If you look at PII data including your full name, your postal address, email address, contact numbers is known by your friends and relatives. Sometimes your mother's maiden name, date of birth and even your passport number and your identity document information may not be confidential to your closer complainants. Let's assume, security question to reset your password is mother's maiden name. If one of your close relatives wanted to reset your password, its super easy as they know about your mother's maiden name. That's the reason, we should not be using accurate information when not required. Specially for the security questions we shouldn't be using our accurate information. This information only required when you need to reset your password if forgotten. The question will be, if you forgotten your password, we should know the answer to the security question to reset it, if you provide inaccurate information how you remember what you provide in your security question. If you think about this in smarter way, probably you can use your partners information as yours when answering security question, so if required you remember whose information you have used.

Best way to protect your credentials is to enable 2FA or **MFA** to protect your credentials. 2FA/MFA uses any of the two from something you know (Password, PIN), something you have (Mobile phone, hardware key) and something you are (biometric authentication).

In today's world many users have multiple account information to remember. When security administrators enforced password complexity policies on the credentials, for the users it's hard to remember. Most of the users use same password for multiple accounts which is not a good practice. As we discussed earlier, if the same password used, attackers can find out multiple account used by the same user and if the one account is compromised, attacker can compromise all the accounts connected to the same user.

Solution is using password manager to keep all the credentials and secure the password manager database using master password. There are many password managers including:

- KeePass
- LastPass
- Bitwarden
- 1Password
- Dashlane
- Keeper Password Manager

The following is an example of KeePass password manager:

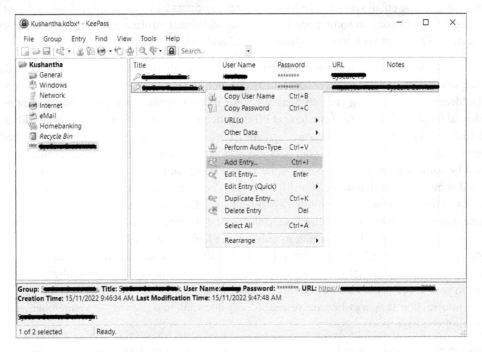

Figure 3.11 – KeePass opensource password manager

KeePass is opensource password manager with secure and stronger encryption algorithms including AES-256 to protect your credentials. Whenever required you can decrypt and copy credentials and paste it on the applications. This can be used as a solution for keyloggers. If you are looking for online solution, you can use Bitwarden, which is opensource online password manager which supports most of the latest browsers. BitWarden uses encrypted vault to keep credentials. BitWarden can be used to create stronger passwords as an additional functionality. BitWarden is free for personnel user with single username, but for professional and enterprise they have separate licenses:

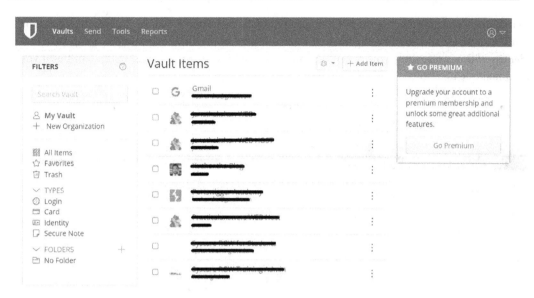

Figure 3.12 – BitWarden open-source online password manager

There are many tools and online services which can generate information that can be used without revealing your real **Personally Identifiable Information** (PII). When you use the information generated by these tools will prevent exposing PII to untrusted entities.

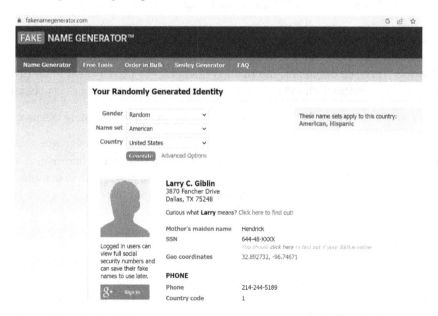

Figure 3.13 – Fake name generator to generate personnel information

Fake Name Generator is an online site that generates fake name and personnel information based on gender, name set and country. It generates interesting set of information that you can create a fake identity without revealing your real PII. Same tool provides option to create bulk identities if you want multiple identities:

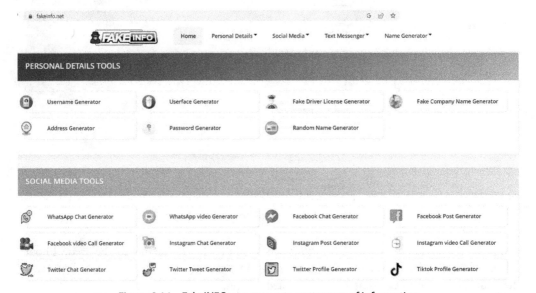

Figure 3.14 – FakeINFO can generate vast range of information

FakeINFO is another online resource where you can generate broad list of personnel information including usernames, addresses, fake driver license, addresses and company names depending on your requirement. Then you can use the information whenever you need to enter in untrustworthy websites, blogs, or any service which will make your personnel information available in public internet.

Also, the same website can be used to generate fake social media outputs including WhatsApp chats, Facebook chat and YouTube channels.

From these sources we can understand that we shouldn't be disclosing our real information always as most of the sources can make our personnel information available in public sources that attackers can obtain quite easily. We should be reducing our footprint on the public internet if you wish to reduce attack surface and the risk on our personnel and sensitive data.

Facts we shouldn't be disclosing

Now it has come to a point where posting, or sharing anything online, sharing by email or social media can introduce new threats. As we discussed before, it can increase the attack surface. It's always good to think before you post or share anything on the internet. Main reason is anything you post, or share can be seeing by anyone. As we discussed in the previous chapters, sharing sensitive and PII is risky and should be avoided. Along with that, consider removing your name and contact information published on websites and public records.

Remove personnel information from WHOIS

Often, we use personnel information in DNS registrations and other public databases. Without our knowledge this information can be collected by attackers using OSINT techniques. Typically, WHOIS records contains registrant contact, administrative contact and technical contact which contains personnel information. We can use privacy controls to mask personnel information from WHOIS.

Image meta data

When sharing photos, meta information attached to photos can go with them including GPS location and other information as we discussed previous chapters. There can be targeted and untargeted social engineering attacks which can collect sensitive information without your knowledge. We shouldn't be sharing images anywhere including social media without removing the meta information embedded to the photos. On the first chapter we discussed how much information can be extracted from an image. We can remove meta information from any image using windows settings or there are many tools available to do that. If you want to remove personnel information from an image, if you are using Windows operating system, you can right click the image and go to properties. Then click on the **Remove properties and Personnel Information** link and you can remove selected attributes, or all meta information form the image:

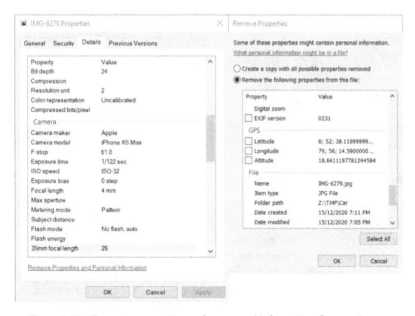

Figure 3.15 – Remove properties and personnel information from an image

Using this option, we can remove properties and personnel information. You can remove selected information or all meta information on an image. This method can be used to remove information form a selected image. There are tools that you can use to remove meta information from bulk set of images at one go. You can download the `Exif Purge` tool from `http://exifpurge.com`:

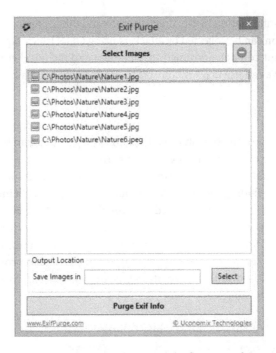

Figure 3.16 – Remove properties and personnel information from set of images

Disposed devices

We often dispose devices or donate devices after years of usage. Sometimes we sell used devices including mobile devices. These devices contain heaps of personnel and sensitive information. We delete data before the disposal. But can we be certain that the data cannot be recovered? There are many tools available to recover data even from deleted partitions.

Commonly used media to store data are hard drives. Typically, digital media stores data as a sequence of 1s and 0s in magnetic media. Data removed by common techniques like data deletion can be easily recovered. Usually, the operating system deletes the reference to the file when file is deleted but the real data never get deleted. Deleted references can be easily created by data recovery tools. This can expose personnel and sensitive data. This has created risk every time we delete confidential data. We think that we deleted data, but in reality, data cannot be deleted, only can over written.

If you really want to remove data to the extent that attackers will never be able to recover, we need to overwrite data completely and multiple times as there can be techniques that attackers use to recover data. Following are some standards for sanitizing media:

- Russian Standard: GOST P50739-95

- German: VSITR

- American: NAVSO P-5239-26 (MFM)

- American: DoD 5220.22-M

- American: NAVSO P-5239-26 (RLL)

- NIST SP 800-88

Driver eraser is one of the tools that is available to securely remove data according to the standards.

Figure 3.17 – BitRaser securely wipe data beyond recovery

Even though data wiping software claims that they remove data securely to unrecoverable state, many organizations still prefer hard disk shredders to physically destroy data.

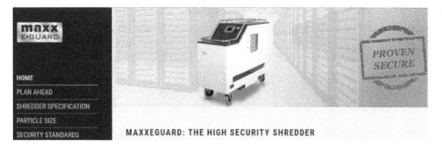

Figure 3.18 – Hard disk shredders to physically destroy data

Hardware shredders physically destroy data by spreading hard drives to small particles where any of the data recovery tool cannot recover. Many cloud service providers use this technology to shred hard drives when they wanted to replace damaged or older drives with new ones as they must comply with compliant standards.

Unsecured cameras

As we discussed in earlier chapters, attackers can use device search engines like SHODAN or Google to find internet connected camera systems. For an example if you search `intitle:"Blue Iris Remote View"` on Google, you can find plenty of cameras that you can view without any credentials. These cameras can be viewed without any credentials. Some cameras installed in common areas, and some are even in private stores. This type of devices can expose personally identifiable information to attackers:

Figure 3.19 – Simple Google dork connects to live camera feed without credentials

When you use google dorks to search, these devices can be accessible by anyone, and any piece of information exposed can be used by these devices can be used by the attackers against you.

Social media matters

Social media has introduced a paradigm shift in communication and interaction with family, friends, and colleagues. Social media has become a part of social life as most of us can interact with our closer circle of friends, relative and colleagues irrespective of the physical and geographical boundaries. Even though social media platforms including Facebook, LinkedIn, Twitter has introduced revolutionary change of communication and introduced many benefits, they also pose significant security and privacy risks to companies, organizations, and individuals. Most of the organizations and individuals are on one or more social media platforms due to their popularity and various benefits.

Concerns on social media

Social media can increase attack surface acting as vulnerable platform for the attackers to launch various types of attacks. Some of the serious concerns when using social media can be as follows:

- **Data privacy and over exposure** – Users often share their information without masking on social media can cause privacy breaches. Most of the social media platforms default privacy setting is *Public*, meaning is anyone on the same platform can view users' information including shared images embedded with metadata.

- **Malicious and third-party applications** - most of the social media platforms allow users to integrate third-party applications. Third party applications can be gaming apps, communicating on apps, location related apps and many more. When the third-party app is integrated, there is a possibility that third-party app can access your data. Most of the apps request your permission to access your information including contacts and location information. But there can be malicious app running in the background can access your personnel information without your knowledge.

- **Data mining** – Most of the social media platforms perform data mining by the information collected during the registration or after the registration including your name, location, age, date of birth and your behavior. This information will be used by the platform for targeted advertising. Even your behavior patterns, can be used for targeted advertising.

- **Malware attacks** – malware attacks can be performed over the social media in multiple ways including shared links, shared files, link redirection, pop-up windows, or advertisements.

- **Impersonation** – Attackers can create fake profile of an internal employee and send friend requests to other employees and can be a part of many important groups to collect information. This can happen to individuals in a similar way, creating fake profile impersonating your friend, then start chatting or through messages collect information and share malicious links. Others trust the links shared as the profile impersonates a known and trustworthy person. Then people assume attacker is internal employee and share confidential information without suspecting. This attack can be performed in WhatsApp groups, Viber groups and Google groups.

Risks involved with social media

When using social media, there are range of risks involved. Social media is a gold mine for scammers. There have been many social media scams through which organization and individuals lost their privacy and collectively lost billions of dollars. There are many common scams, but attackers use different techniques to lure users for the scams:

- **Banking and payment services scam** – scammers will send you a fake banking link or payment service link to perform transaction tricking you to enter your information including bank credentials.

- **Lottery and gift card scam** – users receive *congratulations* message on their email or social media, stating you won a lottery and ask you to enter information including banking information to transfer your winning price. Often attackers might request you to transfer small amount as administration fee to get bigger lottery amount of gift card.

- **Romance scams** – This is a fraudulent act performed by scammer pretending romantic interest in the target. This includes stablishing relationship and maintaining it for some time. After the required level of trust is gained, scammer might attempt to get money for convincing reason or collect sensitive information. Sometimes scammer will send you some gifts or photos to gain high level of trust. Often, they will pretend that they are going to book tickets to come to see you. Also, scammer will pretend that they need some money for personnel emergency like surgery for their mother and try to get funds from the target.

- **Forex scammers** - Often scammers would try to create your interest in forex trading by convincing screenshots and trying to influence you indirectly. Some scammers introduce other fraudulent scammers as forex account managers and try to get your information and money. There are chat groups convincing new users to the group, most of the members are scammers and they try to influence others to invest on fake forex services.

- **Escort scammers** – scammers pretend that they provide escort services and share advertisements on social media like Facebook pages, WhatsApp groups. Often, they advertise on personnel classified pages in web portals like `https://www.locanto.com`, `https://www.skokka.com` with contact information. Mostly they encourage text and WhatsApp messages. When the interested people contacted, they share fake photos, services offered and their rates. Once both parties agreed, scammer will demand a half payment to confirm the booking on payID (payID is a unique identifier linked to attackers mobile or Internet banking) Depending on the country there are similar payment methods that can transfer funds almost instant as mobile cash. When the half payment is made, scammer disappears or block your number.

- **Social media account disabled scam** – scammers send an email to the target, stating that they received a request to disable your account or account is already disabled. To activate your account scammer will request your personnel information or they will provide link to enable your account providing sensitive information. Both ways attacker collects your sensitive information.

Solutions for the risks introduced by social media

Not only scams, but there are many other risks including Cyber stalking, Cyber bullying, and Cyber terrorism exist on the social media. There are solutions to minimize the risks that involved with social media:

- **Protect your credentials** – Create a strong and complex passphrase (Combination of upper case, lower case, special characters, digits and more than eight characters), enable 2FA/MFA if possible and never share passphrase with anyone. Consider using password manager or vault.

- **Treat social media as showcase, everything that you share matters**. Be sensitive on everything you share. Never share personnel information or sensitive information on social media. Think again before you share images. Often images carry more information than we think. We discussed metadata but images may directly carry information in the background. As an example, a couple posted selfie taken during the candlelight dinner may include the credit card they use to settle the bill still on the table or photo taken in the office wearing your office access card can provide enough information for an attacker to duplicate the card.

- **Common sense** – This is the best way to protect yourself from scammers. Most of the scams are too good to be true. Simply, you can't win a lotter that you never entered, or you can't receive a courier that you never ordered. No one will give away hundred iPads to celebrate their anniversary without any gain. You will never receive anything by doing nothing.

- **Reduce the information you disclose over the social media** - Consider using alias than using full name in social media. Consider using separate email address for social media. Try to prevent disclosing your home address, work address, phone numbers, contact information on social media.

- **Never access your social media accounts using untrusted devices or untrusted networks** – never access your accounts by public devices, Common devices in airports and internet cafes.

- **Configure privacy settings** – consider configuring privacy settings, which reduces exposing your information to public. Depending on the social media platform, these settings are very, typically every social media provide an option to configure privacy settings.

- Don't blindly accept friend requests from strangers.

- Don't submit social media surveys require your personnel information even if they claim that there will be a prize for the winners.

- Never download apps via links posted on social media.

- Never share personnel information even if you trust the person as profile can be impersonated.

- Never click suspicious link posted in social media.

Most of the social media platforms available as mobile apps. When downloading app, we need to make sure that the app is downloaded from the trusted app store. Device operating system must be up to date and the App should have the latest update.

Spot potential scams on social media

Most of the scams looks too good to be true. Still the users believe *today can be my lucky day* and falling to common traps. Most of the social media related scams starts with common friend request. Whenever you receive friend request, better to screen the request for:

- If it is newly created profile with row and limited content.

- Many grammatical and spelling errors.

- Content is very generic, not specific.

- Low number of friends or all of them in opposite gender.

- Try to understand their motive by sending you a *friend request* unless you have common friends.

- Profile pic looks like a celebrity or face is not properly visible.

- Be skeptical if suddenly they wanted to something that you are interested, and they request for some amount of money. Probably they wanted to book an air ticket to come to your city and they are short of some amount. Sometimes one of their very close family members need to undergo a surgery.

- If they ask you to add them on WhatsApp or Telegram after short chat.

- If they try to convince profitable business which is very much favorable to you.

- If you receive direct messages asking cash, or gift card or crypto currency for a service, or asking cash advances before the service.

Once you accepted the friend request, they are trying to be very friendly in a very short period. Trying to share personnel pics or video of them and encouraging you to do the same. Think before share anything personnel as often attackers may try to blackmail you.

Summary

This chapter explained overexposure and types of exposing personnel and sensitive data even without our knowledge. Everything that we share or post on the cyberspace can expose our information. Always we don't have to share our real information and there are ways that we can protect privacy without disclosing authentic information. Sometimes we disclose our information that we shouldn't have disclosed. There are ways to reduce the information that we share online. Social media can be a gold mine for attackers to collect sensitive information of targets without disclosing their real identity and how we can protect ourselves from such attacks.

In next chapter you will learn about the techniques that attackers compromise privacy.

4
Techniques that Attackers Use to Compromise Privacy

Newer technologies are introduced frequently, and our lifestyles constantly move us closer to cyberspace. Our relationship with the internet is getting stronger while exposing us more to the unregulated internet. A little over two decades ago, the only way we connected to the internet was through emails. We had dial-up connections, which prevented us from creating a stronger relationship with the internet due to the slow speed and time-based connectivity. When faster broadband connections were introduced, we started forming stronger relationships with cyberspace due to speedy and stable connectivity, irrespective of the time we stayed connected.

In the dial-up days, we connected to the internet only when we really wanted to. Now, we are always connected, whether we want to or not. Previously, we had to put in efforts to connect to the internet; now, we have to put in efforts to disconnect from the internet. The most critical problem, above all, is that while you are connected to your bank account app on a smartphone, another hundred apps on the same phone are connected to the internet using the same connectivity. When this connection is created, a number of layers collect your information, including your device, the **Internet Service Provider** (**ISP**), your browser, and connected apps and websites.

There are multiple ways that attackers compromise your privacy. This chapter concentrates on a range of techniques that attackers use to compromise your privacy, including the following:

- Information gathering
- Enumeration
- Identity exposure
- Artifacts that can be collected from devices
- Social engineering

Information gathering

Hackers often use a range of information-gathering techniques to collect information about a target. This is an important step of hacking, as information gathering provides insight into the target. Typically, there are five phases of attack. These phases are as follows:

1. Information gathering
2. Scanning and enumeration
3. Gaining access
4. Maintaining access
5. Covering tracks

Of the five phases, information gathering is the most important phase, where the attacker spends a considerable amount of time gathering information about the target. If the attacker collects more information, then it's easier to understand the attack surface. Information gathering is usually classified into four categories:

*   Footprinting
*   Scanning
*   Enumeration
*   Reconnaissance

The information that the attacker could collect includes user information, IP addresses, DNS information, subdomains, and company information, which consists of company employees, usernames, open ports and services, operating system versions, and so on. Attackers use a range of technologies and tools during information gathering. During information gathering, attackers often use **open source intelligence (OSINT)**. Open source information is usually collected using major search engines, web pages, and other sources but is not limited to the searchable internet. A considerable portion of the internet cannot be found or searched using major search engines. This is called the **deep web**. The deep web consists of a large number of websites, databases, files, and other resources that cannot be indexed by Google, Bing, Yahoo, or any other commonly used search engines. Attackers use a range of tools, including web spiders, WHOIS databases, Maltego, intercepting proxies, and web resources such as Netcraft.

Types of information that attackers try to discover include the following:

*   Public IP addresses
*   Subdomains and DNS-related information
*   Unintentionally shared sensitive documents

- Directory listings
- Leaked usernames and passwords

Information gathering is the first step of hacking and ethical hacking. Information gathering is a kind of art that any attacker or penetration tester should learn and master. According to the method that the attacker uses, information gathering is divided into two types:

- Passive information gathering
- Active information gathering

Passive information gathering

Passive information gathering is a method of gathering information by using other available sources without connecting to the target. When performing passive information gathering, attackers use search engines such as Google. This method is commonly known as **Google hacking or Google dorking**. In other words, attackers use Google's capabilities to find information.

Attackers need to know published and unpublished subdomains. They use a range of tools to discover subdomains. You can use online services to find out this information – for example, `https://searchdns.netcraft.com/`.

Figure 4.1 – A Netcraft search for subdomains

Attackers can use online tools to search subdomains of websites, using tools such as Netcraft. There are many tools that can be used for this purpose, including the following:

- **Sublist3r**: A subdomain scanner used by bounty hunters
- **AMASS**: A subdomain discovery tool

- **SubBrute**: A very fast subdomain brute-forcing tool

- **Knock**: A subdomain scanner written in Python

- **DNSRecon**: A collection of domain reconnaissance scripts written in Python

- **Altdns**: A permutation-based subdomain discovery tool

When you search the URL on Netcraft, it will generate a list of subdomains of the main domain. These subdomains can provide very interesting information, as some of them may not be exposed to the external world. These subdomains can be used for testing, internal use, and installing web apps. Netcraft can also generate DNS reports, which contain interesting information about the target, including the range of the IPs that were used by the URL, historical data, and operating systems used to host a web application.

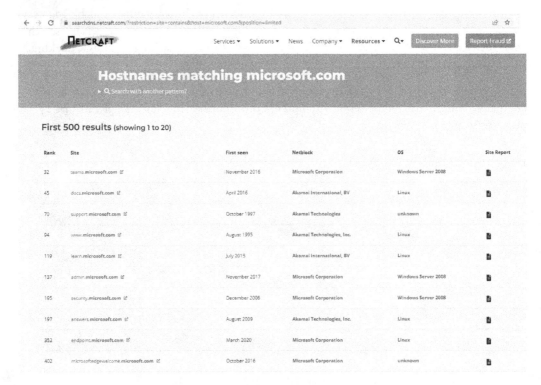

Figure 4.2 – A Netcraft search for subdomains on Microsoft

When a company or an individual shares information on their own website or social sites, this information can be used by attackers. Hence, we need to know to what extent we should be sharing information on the web. Exposing more information than required is known as information overexposure.

Even the disclosure of a subdomain can create an entry point for an attacker. Sometimes, administrators may install content management systems in a subdomain for testing. It is possible that they might still be configured with default credentials or default installation. Once an attacker has identified the subdomain and scanned it, they can use the default configuration to reconfigure and gain access to the server or default credentials.

WHOIS lookup

WHOIS is a query-based protocol that is used to query information stored in a database related to users of internet resources. This is a publicly available database that contains information such as assigned domain names and IP addresses. Often, attackers use this to gather information and can use the `nslookup` tool to obtain DNS entries of the domain or online tools:

```
C:/nslookup
Set q=any
Microsoft.com
```

The Wayback Machine

The Wayback Machine is an internet archive that keeps over 650 billion web pages as a digital library, which allows users to go back in time and explore how websites looked in the past. This is a not-for-profit organization that keeps snapshots of websites around the globe, not just famous ones but also all possible websites. You can access the Wayback Machine at `http://web.archive.org/` and search any URL you want. For example, if we search `www.yahoo.com`, it will show us the snapshot that the Wayback Machine has in its digital library.

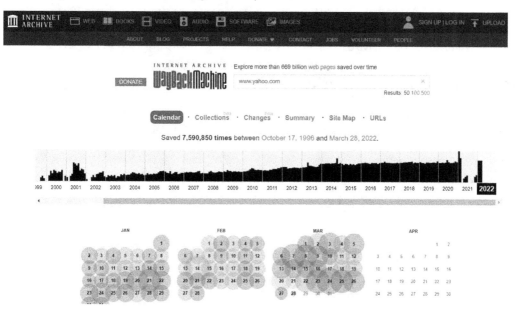

Figure 4.3 – Wayback Machine snapshots of www.yahoo.com

When you click any of the snapshots, it will take you to the web page that is stored in the digital library. You can even search for your own websites to see how many snapshots are saved in the digital library. If we select the www.yahoo.com snapshot that the Wayback Machine took in February 1997, it takes us to the relevant web page.

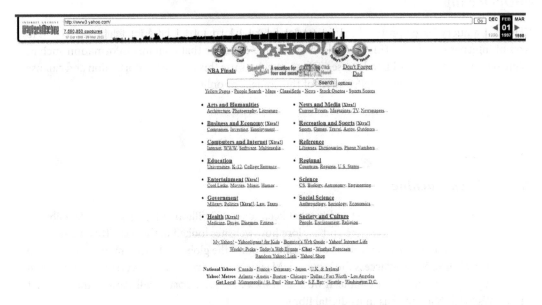

Figure 4.4 – A snapshot taken by the Wayback Machine of www.yahoo.com in 1997

These repositories may still have data that you deleted long ago.

Active information gathering

Information gathering by connecting to a target is known as active information gathering. Active information gathering can be used to discover useful information about the target, including the following:

- Open ports and services
- SMB shares
- IP addresses and address shares
- Operating system versions and applications
- Information related to infrastructure

There is a range of tools that an attacker can use for active information gathering. Nmap is the most used information-gathering tool by attackers. It is a free command-line tool, but a GUI-based version can also be downloaded. Once it has scanned a target, nmap returns information related to the range of IPs or the specific IP address that the attacker needs to gather information on.

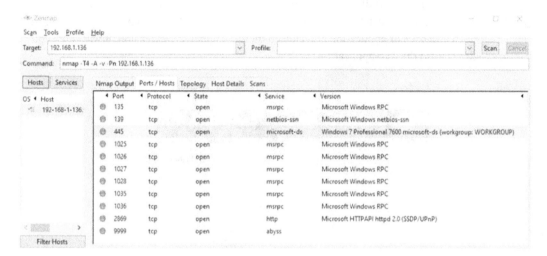

Figure 4.5 – An nmap scan shows open ports of a target system

This result generated after an nmap scan shows the list of open ports, services running, and service versions of the target system, which provide useful information to an attacker to uncover vulnerabilities. Depending on the scope, attackers use different tools. nmap is mostly used to carry out internal or external network-related scanning. If the attacker wants to scan a target website, there is a range of tools they can use.

Any type of information that we may expose can be used by an attacker. By reducing the amount of information that we share, we can reduce the attack surface. This information can be exposed by a company website, vacancy advertisements, and company portfolios, among others. For example, if a company wants to recruit an IT manager or systems administrator, they often advertise what they expect from candidates, including technical capabilities and work experience on specific products. This requirement can expose company infrastructure, including the products that the company uses within its IT infrastructure. Then, attackers can understand from the information on the vacancy advertisement the email server version of the company, the firewall model, the size of the infrastructure, and other security mechanisms in place. Some employers even share this information on their social media sites, such as LinkedIn, when they describe their job roles, thus often overexposing the company's IT infrastructure and sensitive data.

If the scope is a web application, the risk is even higher, as web applications are published online and accessible over the internet. Attackers use a range of tools to scan and understand web applications. There are online scanners and offline scanners. There are some browser-based extensions used by attackers to collect information about web applications. The *Wappalyzer technology profiler* is such an extension, which can be installed in the Chrome browser, and attackers can analyze the web application on the go. You can install Wappalyzer by searching for it in the Chrome extensions. Once you have accessed a website or application and clicked on the Wappalyzer extension, it will show you the technologies behind the web application.

Figure 4.6 – Wappalyzer shows technologies used by the web application

There is a range of other tools used by attackers to collect information about targets.

Enumeration

Enumeration is a process during information gathering where attackers gather very specific information. Unlike passive information gathering, during enumeration, an attacker actively connects to a target and collects specific information. Since the attacker is connected to the target, the attacker can send direct queries to the target to collect specific information, which is useful to exploit a system to gain system access. During the enumeration phase, attackers will try to gather information such as the following:

- Users and group names
- Network and file shares
- Computer names
- Application-related information or banners
- Routing and **Address Resolution Protocol (ARP)** tables
- DNS information
- User lists from web applications and **Active Directory (AD)**

Enumeration is an important phase and attackers use a range of techniques to enumerate targets. Depending on the target, attackers change the technique as well. The most common enumeration techniques are as follows:

- Extracting information using default passwords and word lists
- Brute-forcing an AD
- NetBIOS enumeration
- User and group extraction in Windows
- Extracting DNS information
- Printer shares

Attackers specifically target a range of ports and services for enumeration. These ports and protocols are prone to enumeration attacks. The following are the most common services and ports targeted by attackers:

- TCP 445 SMB
- TCP 139 NetBIOS
- TCP 389 LDAP
- TCP 53 DNS
- TCP 25 SMTP
- TCP 135 Microsoft RPC

Utilities and tools are specifically targeted on these services and ports to extract information during enumeration.

Depending on the target, attackers use different tools to collect enumeration information. For example, attackers use `https://dnsdumpster.com/` to collect web-based application targets.

Figure 4.7 – DNSdumpster shows information on the URL that is given

When attackers scan the targeted web URL on this portal, it generates very important information about DNS servers, MX records, TXT records, host records, and domain maps. They can collect valuable information such as the following:

- **DNS servers**: Collecting information from the DNS servers and banner information.
- **MX records**: Mail exchanger records and hosts sharing the same IP.
- **TXT records**: All text records created on a domain.
- **Host records**: All host records and banners.
- **Domain mapping**: This has all the connected hosts and domain information of a main domain. This reveals information related to the domain and connected hosts and presents it in a graphical view. The domain map also shows the interconnection between domains, subdomains, and domain records.

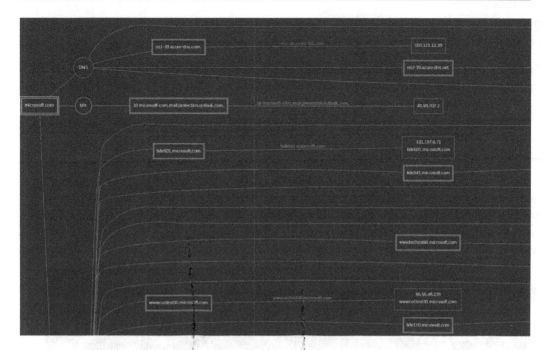

Figure 4.8 – The domain map shows domain information

Depending on the target, attackers use multiple methods to collect information. As an example, if the target is a website, attackers try to find out legitimate usernames of the target website. WPScan is a tool that can be used to enumerate the username of a WordPress website. WPScan is a free and open source tool included in Kali Linux and many offensive security distros.

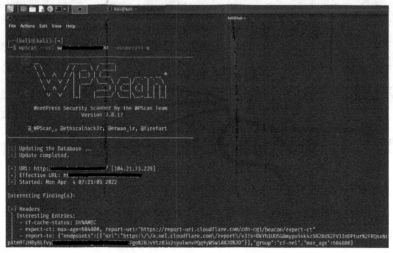

Figure 4.9 – WPScan enumerates usernames from a WordPress website

Once scanned, WPScan enumerates the list of users in the WordPress web application. Attackers can use collected usernames during the gaining access phase to compromise the systems.

Gathering information from complex networks with thousands of devices is not easy. Attackers use enumeration techniques to gather information about the connected devices from NetBIOS caches. A NetBIOS cache keeps information about recently resolved IP addresses. Rather than spending hours scanning complex and large networks, this is an ideal solution to collect accurate information readily available in a NetBIOS cache.

If an attacker is connected to a domain environment, enumerating usernames and groups is another common way of gathering specific information related to targets. In most enterprise networks, it is common to use AD as an identity system, if they use Microsoft technologies. When AD is used to maintain authentication and authorization, username and groups are created in an AD. AD enumeration tools are used to enumerate users and group names from the AD. Enum4linux is a commonly used AD enumeration tool. If the attacker has access to the network, they can run this tool to collect information.

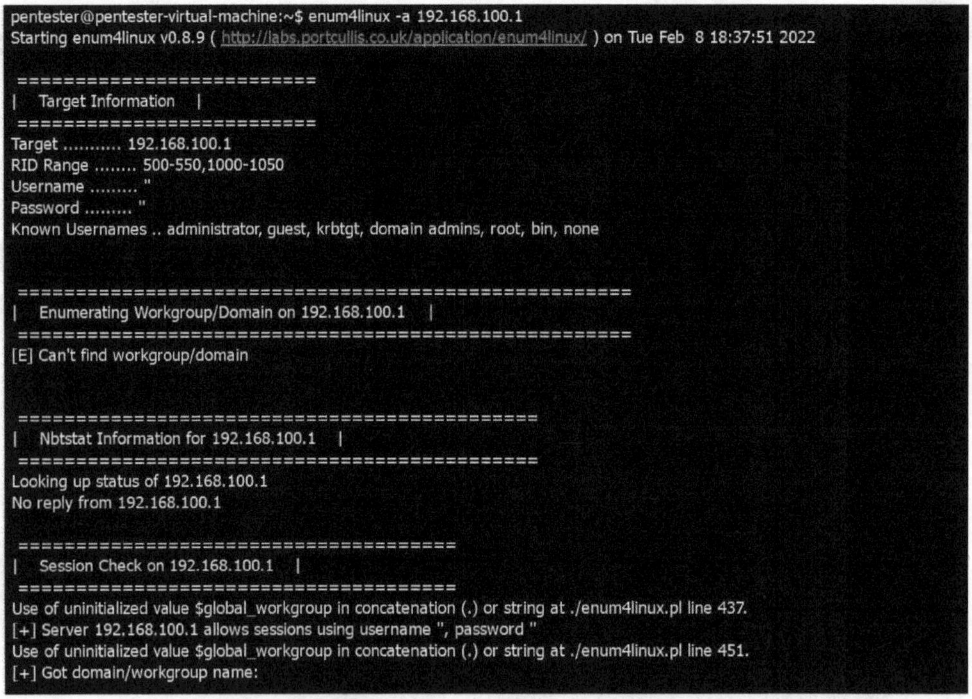

```
pentester@pentester-virtual-machine:~$ enum4linux -a 192.168.100.1
Starting enum4linux v0.8.9 ( http://labs.portcullis.co.uk/application/enum4linux/ ) on Tue Feb  8 18:37:51 2022

 ==========================
|   Target Information   |
 ==========================
Target ........... 192.168.100.1
RID Range ........ 500-550,1000-1050
Username ......... "
Password ......... "
Known Usernames .. administrator, guest, krbtgt, domain admins, root, bin, none


 =====================================================
|   Enumerating Workgroup/Domain on 192.168.100.1   |
 =====================================================
[E] Can't find workgroup/domain


 ============================================
|   Nbtstat Information for 192.168.100.1   |
 ============================================
Looking up status of 192.168.100.1
No reply from 192.168.100.1


 ===================================
|   Session Check on 192.168.100.1   |
 ===================================
Use of uninitialized value $global_workgroup in concatenation (.) or string at ./enum4linux.pl line 437.
[+] Server 192.168.100.1 allows sessions using username '', password "
Use of uninitialized value $global_workgroup in concatenation (.) or string at ./enum4linux.pl line 451.
[+] Got domain/workgroup name:
```

Figure 4.10 – Enum4linux enumerates usernames from AD

When the attacker is connected to the network, they can use the enum4linux tool to enumerate the users and groups.

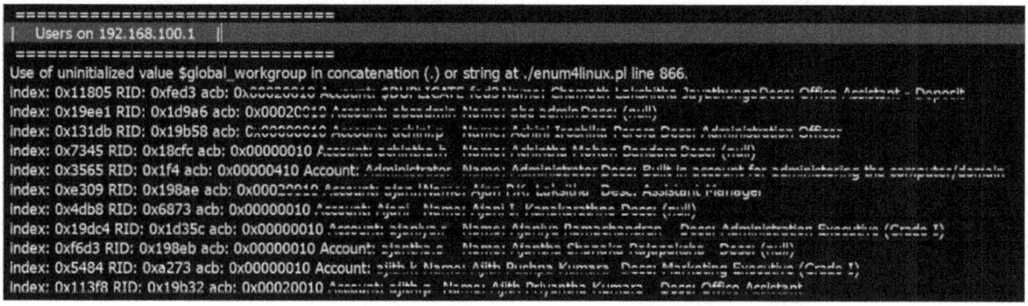

Figure 4.11 – Enum4linux enumerated usernames from AD

The preceding screenshot shows that enum4linux was used to acquire a list of users and their information from AD.

Identity exposure

An individual using multiple devices has become the norm in today's world. Most of the devices we use today store our identities in various forms. This allows our identity information to be easily exposed. This can happen at an operating system level, an application level, a storage level, or a network level. Using various devices makes our life easy, but we are sacrificing identities and other sensitive information by ignoring privacy. If a device is stolen, stored data can be accessed easily unless the device storage is encrypted. Most identities are exposed due to unawareness of various types of attacks.

Individual identities can be exposed in multiple ways, including the following:

- Publicly available sources
- Blogposts and registrations to newsletters by email addresses
- Social media
- Registrations to various services using identity information
- Selfies and other images published on different platforms
- Online and offline forms
- Using shared devices and sold and stolen devices

Attackers can use the preceding methods to discover your identity without your knowledge.

It's not only individuals but also enterprises that suffer from identity exposure. Organizations use enterprise identity systems such as AD to authenticate and authorize user entities to services and resources. Most organizations don't have the required visibility of identities and their exposure at their endpoints. Most importantly, identities can increase an attack surface. When an organization doesn't have the required visibility of its identities, it is difficult to identify the risks introduced by identity systems. Identities can introduce deadly exposures, misconfigurations, and incorrect permission allocations, which can increase the attack surface to a great extent.

Modern IT systems have many forms of identities, including the following:

- **On-premises identities**: For example, an AD

- **Cloud identities**: For example, Azure AD

- **Hybrid identities and linked identities**: For example, an AD and Azure AD linked by AD Connect

When enterprises deal with identities, they must always ensure identities are not exposed. This can be performed by adopting an attacker's point of view. This can be done through vulnerability assessments and penetration testing.

Identity exposure has led to identity-based attacks, including account takeover attacks and a range of password attacks. Preventing these attacks is increasingly difficult as new ways of identity exposure are introduced on a daily basis. Attackers collect personal and sensitive information from various sources, including via credential leaks, social media, breachers, and publicly available information, to launch these attacks.

Identity attacks range from common password-related attacks, such as wordlist attacks and brute-forcing, to common user password profiling. Once an identity is exposed, attackers can easily build wordlists with usernames. Large password wordlists are usually available to download from multiple sources, including password generators. There are two main types of password cracking:

- Online password cracking

- Offline password cracking

Online password cracking

Online password cracking is a method that attackers use to compromise the password of online services, including those beginning with http, https, ftp, and smtp, by using attacking tools and probing online services with the help of a password list. You can download very large password lists directly from the internet by just googling. Once, a user in a hacker forum leaked 8.4 billion entries of passwords in text format in a massive 100 GB .txt file. This file contained passwords between 6–20 characters long without duplicates or spaces. Typically, attackers use two methods in online password hacking: **dictionary attack** and **brute-forcing**. A dictionary attack makes use of a password file known as a dictionary and tries all the passwords one by one until the password matches or you've reached the end of the file. The brute-force attack, meanwhile, tries all possible passwords generated from a given character set. The dictionary attack is a better choice over the brute-force one for online password cracking, considering the time that is required for the brute-force method. Brute-forcing is always successful if the required time and character set are available.

If the identities are exposed, attackers can always try online password attacks, especially for online services, including web applications. A credential breach is one of the most common and most used ways to breach systems. Attackers are interested in credential breaching, as it gives them complete control over a target. Most web applications are designed to provide full access or functionalities for legitimate users. If the attacker manages to get the correct credential from the online password cracking, that allows them to make use of any kind of capability offered to legitimate users.

Another important aspect of online password cracking is the vast range of protocols that accept online usernames and passwords to authenticate, and that can, therefore, be attacked by online password cracking from anywhere in the world over the internet. Attackers can strike from anywhere in the world if a service or protocol is open to the internet and accepts a username and password.

One of the disadvantages of online password cracking is it is very slow and has many dependencies, including network speed and service responses. There is a range of password cracking tools that support online password cracking, including the following:

- **Hydra**: One of the most used free online password cracking tools that supports a range of protocols

- **Medusa**: A faster tool when compared to Hydra but fewer services and protocols supported than Hydra

- **Patator**: A Python-based password cracker-supported range of protocols

Offline password cracking

Offline password cracking involves attempting to recover passwords from a password file. Typically, passwords are stored in an encrypted form. When you save a password or create one, it will be saved as a hash. As an example, if you enter a password as `P@$$w0rd@123`, it will be stored as `931e25cb35290d1eaff3a42f03f8d1f8` if the hashing algorithm is MD5. Depending on the hashing algorithm used in the application, the output will be different. Every hashing algorithm generates a fixed-size output, irrespective of the type or size of the input. Hashing is also irreversible.

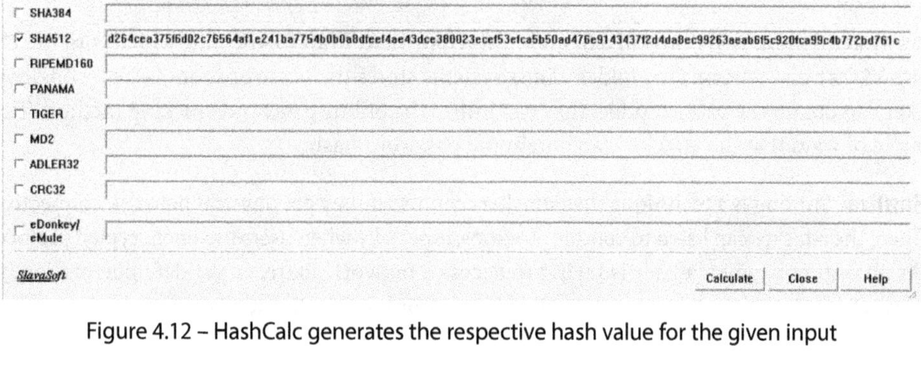

Figure 4.12 – HashCalc generates the respective hash value for the given input

Since hashing is irreversible, offline password cracking is a two-step process. If attackers use wordlists for offline password cracking, the steps to recover the password will be the following:

1.   Take the first word from the wordlist and generate the hash using the respective algorithm.

2.   Compare the hashed output with the value in the password file.

If the two values match, the password is the word, and if not, the same process should be continued with the next available words in the word file, also known as the dictionary.

However, some tools and applications claim that they can crack a hash. As an example, we can use `https://crackstation.net/`. CrackStation supports a range of algorithms, including LM, NTLM, md2, md4, md5, md5-half, sha1, sha224, and sha256. CrackStation does not crack a hash but compares the entered hash with a long list of precomputed cracked password hashes. If CrackStation can locate a hash, hash that is like the entered hash value, it returns the respective plain text password. If the same hash value that we generated using HashCalc is entered into CrackStation, it will return the respective plain text value, as seen in *Figure 4.13*:

Figure 4.13 – CrackStation recovered the plaintext password from the entered hash input

Windows systems store passwords in a **Sequential Alignment Map (SAM)** file, which is in the `C:\WINDOWS\System32\config` folder. Linux systems stores the passwords on `/etc/shadow`. If an attacker can obtain the password file, they can utilize the offline password cracking method. There are a couple of ways that the attacker can obtain the password hash:

- **Sniffing**: Sniffing is a technique that attackers can use if they get physical network connectivity. Then, the attacker can listen to communications, especially when users use unencrypted protocols. As an example, when a user is trying to access a network share, to validate permission, the network share will challenge the user for the password. If the attacker listens to the network and collects the challenge and the response, the attacker can obtain the password hash. Attackers use tools such as Wireshark, Ettercap, Cain and Abel, and tcpdump for sniffing.

Figure 4.14 – Sniffing plaintext credentials of a user login using the HTTP protocol

- **Physical access**: If an attacker gets physical access to a device, they can obtain a SAM file. Typically, attackers use Linux live systems to boot a targeted system or some other tools, such as *Hiren's BootCD*, freely downloadable from `https://www.hirensbootcd.org/`, which has a long list of tools that can be used to even reset an administrator password.

Artifacts that can be collected from devices

Most enterprise systems provide services through endpoints. Individuals access systems too through endpoints. When you access any system, including a cloud system, an on-premises system, or even an application, there are many artifacts collected by a device. If the device is lost or stolen, a third person can have direct access to these artifacts, unless the device is encrypted as per industry standards. During investigations, forensic investigators can use forensic tools to recover many artifacts and obtain information.

If an attacker has access to a device in the event of it being lost or stolen, they can collect the following information from it:

- Contact information and the phone book can be exported or downloaded.
- Messages, including SMS and application-based text messages.
- Multimedia content, including pictures, videos, and sensitive multimedia content.
- Call history, including incoming and outgoing calls.
- Stored passwords and codes.
- Wi-Fi network association data.

- System files, log files, and system-created temporary files.

- Geolocation information.

- Documents created and downloaded

- Plans and to-do lists.

- Internet browsing history, temporary files, and cookie information.

- Data collected by various applications that are installed.

- Deleted files and artifacts from the system, which can be recovered.

When an attacker has access to all this information, they can use it to do direct attacks and indirect attacks. Types of artifacts that can be collected can vary, depending on the form factor of the device and the operating system. Currently, many devices used in enterprises are Windows, Mac, Android, and iOS. Depending on the device's operating system, attackers can use various techniques and tools to collect artifacts and information from it. If the attacker has full physical access to the device, the number of artifacts that can be collected is much higher. The attacker can even create a copy of the device using physical or logical acquisition methods. Once a copy is created, the attacker can then analyze the device's contents further and acquire more information. This operation is similar to the methods that are used in forensic investigations.

To maintain cyber anonymity, the devices that we use to access the internet are very important. As we discussed earlier, attackers can access sensitive information on your devices, even if they aren't stolen or lost, via applications, malware, and physical device access.

Figure 4.15 – USB Rubber Ducky injects keystrokes at superhuman speeds

USB Rubber Ducky is designed by Hak5 and looks like a USB stick. It can inject a series of keystrokes at superhuman speeds. Even if the USB storage port is disabled in the device, USB Rubber Ducky can still send a pre-created payload by posing as a keyboard. Other similar devices were introduced by Hak5. They introduced this device in 2010 with a simplified *duckyScript* language, which attackers can use to design their own payloads. Conversely, Hak5 itself provides a range of pre-crafted payloads that can be used directly with the device. USB Rubber Ducky was opened to the community when the tool got a lot of attention and many people were sharing creative payloads, with individuals submitting their own payloads in various categories, including credentials, execution, exfiltration, recon, and remote access.

This simple hacking tool can be used by attackers for different devices, ranging from Windows and iOS to Android and Linux operating systems, as almost any device that accepts inputs from keyboards can be emulated by USB Rubber Ducky.

If you keep your device unattended or let an attacker access it, intentionally or unintentionally, for a few seconds, that would be more than enough time for them to connect USB Rubber Ducky to your device and upload the payload to your device or steal all the saved credentials in the device, as all the keystrokes are pre-defined in the payload.

As we discussed in previous chapters, some applications such as *Cerberus* or backdoor tools such as *AndroRAT* can be used to collect information from your device, including contact information, images stored in it, call information, text messages, and social media interactions, including private messages.

There is a range of tools used by forensic investigators that can also be utilized by attackers to access a device or create a complete copy of it, which can be analyzed by the attackers later to acquire sensitive information. Forensic investigators extract data from mobile devices using three methods. These methods are physical acquisition, logical acquisition, and filesystem acquisition. Forensic investigators prefer the first method, as it gives them a complete replica of the device, and a successful physical acquisition provides stronger digital evidence for the investigators.

Attackers don't really care about the forensic value of the acquired data if it provides them with the sensitive information they need. Forensic value entails admissibility in court, which is required in order to be accepted as evidence for any case. Attackers' objectives are very different from that of investigators, even though they use similar tools to extract data.

The following list of tools is freely available on the internet for investigators and can be used by attackers to extract data:

- **FTK Imager Lite**: This tool can be used to create an image of a target device, which can later be used by attackers to analyze the image.

- **Andriller**: This tool allows attackers to acquire information from various social media app databases. This utility can crack lock-screen patterns or PINs and decode some databases used for various communications.

- **Android Data Extractor Lite**: This is a Python-based tool, used to acquire information from Android-created databases.

- **Linux Memory Extractor**: This tool allows an attacker to create a volatile memory dump (acquiring data in a memory) from a target device, and they can then execute this tool across the local area network.

- **AFLogical OSE**: This tool must be installed on a targeted device as an APK, and then an attacker can extract various information from the SD card, including the call log, the contact list, and text messages.

Social engineering

Social engineering is a type of attack in which attackers target users directly, rather than trying to compromise complex systems. It's a psychological manipulation of human nature, which involves tricking them into disclosing sensitive information. An attacker first collects information and investigates a target to find out a potential point of entry and weak security procedures in order to launch attacks.

Types of social engineering

There are two types of social engineering: computer- and mobile-based social engineering and human-based social engineering.

Computer- and mobile-based social engineering

There is a range of social engineering attacks based on computers, including the following:

- **Phishing and spear phishing**: Phishing is a type of social engineering attack where an attacker sends a malicious link to a cloned website or an email that tricks the user or users to enter sensitive information. If the attack is aimed at a specific user or an organization, then it is referred to as spear phishing.

- **Hoax letters**: Typically, hoax letters are meant to scare or warn you about nonexistent malware, viruses, or ransomware, suggesting that you install fake antivirus from a given link. Another type of hoax letter trick to defraud a user is to an email asking you to send money for a particular reason. The reason can be religious, personal, or economical.

- **Chain letters**: This is a type of social engineering attack that involves tricking people into forwarding an email to multiple people in order to get a higher level of reach. The email can contain false information or be malicious.

- **Spam messages**: Spam messages are typically not harmful but irritating. This type of unsolicited message can try to gather information, such as by collecting authentic email addresses.

- **Targeted chatting**: Using instant chat services, including mobile apps and social media apps, attackers may try to chat via fake accounts to gather information. Scripts and bots are developed to automate this process. Some of the bots are smart, as they are backed by artificial intelligence and machine learning technologies.

- **SMS**: Attackers often use fake SMS messages, claiming you won the lottery, your shipment is on the way, your courier has delivered your parcel, or you're selected for a raffle draw, and requesting you to register with confidential information.

- **Malicious apps**: Applications downloaded from untrusted sources or shared links can be malicious. These applications can collect confidential and personal information in the background and share it with attackers.

- **Baiting**: In this attack, attackers attract users with information that the users believe can be useful to have. It can be presented as an important software or feature update; sometimes, attackers can leave USB memory sticks with an interesting label unattended to attract a target.

- **Quid pro quo**: This attack is similar to baiting attacks, but instead of baiting you with a *thing* that is valuable, attackers perform an action. For example, the attacker might call a company and pretend to be technical support returning a call and providing instructions to the targeted user to download and execute pre-created malicious code or script, which provides attackers with access to the target systems.

- **Scareware**: This attack displays a warning notice to the target users, typically something like *Your computer is under attack* or *Your device is infected*. Sometimes, these tools display that they are scanning the target computer and convince users to install antimalware from an attacker's link, which will end up installing the attacker's tools on the target system.

Human-based social engineering

There is a range of social engineering attacks targeting humans, including the following:

- **Impersonating users**: This involves pretending to be a legitimate user and trying to get access to important information. This type of attack can be done over the phone, in person, or by using chat services. Often, attackers pose as a higher authority and ask for sensitive information.

- **Dumpster diving**: This involves collecting information from dustbins or discarded computer systems through data-recovery techniques. This is the reason we should be properly shredding confidential information before disposal. For example, if you need to dispose of a credit card statement or any letter sent to your name and address, you should remove personally identifiable data and sensitive information before disposal. Even if you want to dispose of computer systems, you need to permanently remove electronic data as per data destruction standards, such as NIST 800-88 or DoD 5220.22-M. If the data is recoverable, attackers can use simple tools to recover deleted data, including on mobile devices. Attackers can even recover deleted partitions using simple tools.

- **Shoulder surfing**: This can be performed by an attacker as part of direct observation, such as looking over a user's shoulder when entering important information such as credentials. Attackers can also use surveillance systems such as **closed-circuit television** (**CCTV**) to obtain this information.

- **Piggybacking**: An attacker can pretend to be an employee and request a legitimate employee to allow them access to restricted areas by providing a convincing reason.

- **Eavesdropping**: This involves unauthorized listening to conversations in person, over the phone, or by using similar technology to collect information.

These methods clearly show that we shouldn't be disclosing our information from publicly available sources. Also, we should be aware of social engineering attacks, as there is no other way of preventing social engineering attacks than building awareness. We will be discussing countermeasures for social engineering attacks in the latter part of this chapter.

Tools used by attackers to launch social engineering attacks

User awareness is the key to protecting any entity from social engineering attacks. There is a range of tools available, both open source and commercial, to test social engineering readiness. You can use these tools to test user readiness while attackers use the same set of tools to launch real attacks:

- **Social-Engineer Toolkit** (**SET**): SET is an open source tool written in Python to conduct social engineering penetration testing. Anyone can create and clone a phishing website using SET to launch a social engineering attack. Attackers use a range of tools, including free ones such as SET, to clone websites to lure users and collect confidential information.

  SET comes with pre-created templates that help a user or attacker launch an attack in no time. The built-in web server in SET provides a ready-made web server to host a cloned website. This capability is useful to launch exploits that compromise most web browsers.

  Attackers can use built-in templates in SET to clone a legitimate website, which creates a more realistic output. Also, SET has readily available pre-created web pages for popular web applications such as Google, Yahoo, Twitter, and Facebook.

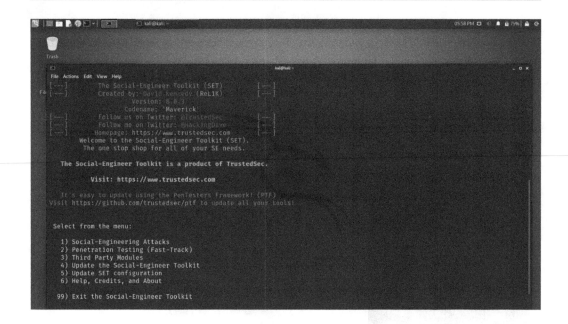

Figure 4.16 – Using SET to create phishing websites

Most tiger boxes, including Kali Linux, have SET built in, and it is also downloadable from GitHub.

When you execute SET from Kali Linux and select **Social-Engineering Attacks** (*Figure 4.16*), it provides a range of options to launch different attacks (shown in *Figure 4.17*).

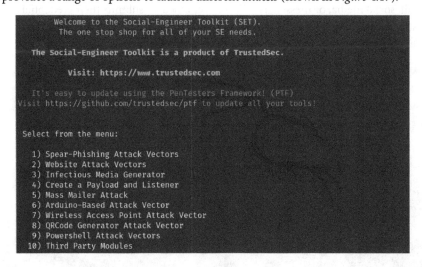

Figure 4.17 – An SET-supported attack list

The most used option by attackers is **Website Attack Vectors** (*Figure 4.17*), after which they can clone any website login page they desire. They can also use built-in templates. On selecting the second option, which is **Google** (*Figure 4.18*), SET uses a pre-created Google template and hosts it on a built-in web server. The attacker can use any IP associated with Kali Linux or use DNS servers to provide a more convincing URL to publish the web page.

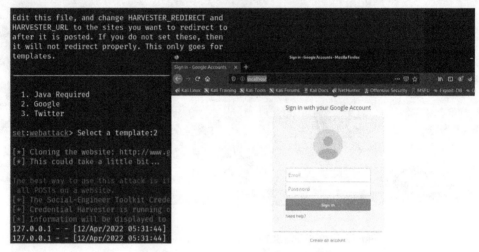

Figure 4.18 – An SET-created Google login

Once an attacker has created a cloned login page or used an existing template, it will be published automatically by a built-in web server. Then, the attacker can share the link with targeted users by email, social media, or messengers. Whenever any user submits their credentials, the real-time attacker will receive the credentials in the SET console or they can store the credentials on a text file created in SET.

Figure 4.19 – SET displays the credentials entered by the target user

Figure 4.19 shows the credentials and other parameters that the browser intended to send to Google servers, which were captured by SET. Once the attacker receives the information, the attacker can log in to the web application using the real credentials that the user entered.

We have only displayed *web attack vectors*, but there is a long list of attacks possible with SET, including more sophisticated attacks. In some complex attacks, attackers can use SET to launch social engineering attacks using payloads created by the Metasploit framework.

- **Wifiphisher**: This tool is built into Kali Linux, or you can download it for free from GitHub. It is used by attackers to automate phishing attacks on Wi-Fi networks to obtain Wi-Fi passwords. Using this tool, attackers can jam nearby Wi-Fi **Service Set Identifiers (SSIDs)**, commonly referred to as Wi-Fi networks or hotspots, and clone them. The newly cloned fake SSID does not require a password to log in. This act is also known as an *evil twin*, as there are two access points with the same SSID in the same range, but users will only see them as a single Wi-Fi hotspot.

  When anyone connects to the fake SSID, there will be a legitimate-looking password prompt, requesting users to enter their Wi-Fi password for a firmware update. The notification further states that without the firmware update, the Wi-Fi will not work. Once the legitimate user enters the Wi-Fi password, it will present the user with a fake update process and reboot timer while notifying the attacker of the Wi-Fi credentials. It's a useful tool for wireless penetration testing and is also used by attackers.

- **Maltego**: This tool is useful for gathering information before launching a social engineering attack. There are three phases involved in a social engineering attack: research, planning, and execution. Attackers must carefully plan the attack first. A successful attack requires proper research and planning. This requires a considerable amount of information gathering to understand the organizational structure, and information about individuals, behaviors, and interests.

  Maltego is an OSINT investigation tool that generates a graphical representation of how pieces of information are linked with each other. We will discuss Maltego in more detail in *Chapter 5, Tools and Techniques That Attackers Use*. Maltego supports a range of inputs, including DNS information, domain names, organization names, and individual names, and then collects information automatically from the integrated publicly available sources, providing a graphical representation of the interlinked information.

There are many tools used by attackers to launch social engineering attacks. Many organizations now use various platforms to test the social engineering readiness of the organization. Building user awareness is the key to preventing many social engineering attacks. There is a list of platforms available for individuals and organizations to test their readiness:

- **Gophish**: This is an open source platform supported by many operating systems. This tool is very user-friendly. You just need to extract the compressed folders and can then start testing right away. Users can be added one by one or by importing CSV files. There, built-in templates can be used for testing by creating campaigns. Reporting is built in, where testers can test the statuses and detect the users who fall for social engineering attacks.

- **Infosec IQ**: Infosec IQ is used to test sophisticated simulations for an entire organization and contains a library of thousands of pre-created templates for testing. This tool also provides a free phishing risk test, which can be used by organizations to launch a phishing campaign and find out an organization's phishing rate.

- **Attack Simulator**: This tool is a part of a Microsoft 365 Enterprise edition subscription or can be subscribed to as an individual feature with a Microsoft Enterprise Mobility and Security subscription. Organizations that have licenses for Attack Simulator can use hundreds of built-in templates to simulate phishing attacks for selected users or all of their users. This also has an interactive awareness education feature built in for users to gain more knowledge about phishing attacks.

- **Simple Phishing Toolkit**: This tool also combines a simulated phishing attack and awareness education for targeted users.

Apart from these tools, there are standalone utilities, such as **SuperPhisher**, which clones any login page in seconds, along with a `.php` file that is capable of writing submitted credentials on a text file, which attackers can access later.

Preventing social engineering attacks

Social engineering attacks have become very sophisticated, and prevention is not that easy. As we discussed in this chapter, social engineering attacks cannot be prevented by security appliances such as firewalls or intrusion prevention systems, as the targets of social engineering attacks are humans. There is a common quote among security professionals – *There is no patch for human stupidity*, which means social engineering attacks cannot be prevented by applying patches or hardening systems. When attackers cannot find vulnerabilities in a target system, attackers can create vulnerabilities using social engineering. It is, therefore, important to prevent social engineering attacks using the following measures:

- **Security awareness and training**: There is a myth stating that *security training is only for IT professionals*, which is *incorrect*. IT professionals represent a very low percentage of the total employees of an organization. Cybersecurity risks are present for every employee who is a part of the system and connected to the internet. We cannot expect IT professionals or IT security professionals to protect every security aspect of an organization. IT staff and security staff can be critical in this, but every employee should hold equal responsibility in cybersecurity. Due to this reason, cybersecurity training should be an ongoing activity at any company.

- **Security products**: Antivirus systems and endpoint protection tools cannot protect users directly from social engineering attacks but can be useful to detect the payloads that can be downloaded as a part of such attacks. Many phishing messages and links try to download or upload malicious payloads to a system; however, if antivirus protection is active, malicious payloads can be eliminated.

- **IPS and SIEM**: Some **Intrusion Prevention Systems (IPSs)** and **Security Information and Event Management (SIEM)** solutions have features such as **User Event Behavior Analysis (UEBA)**, which creates a baseline for user behavior that is capable of detecting anomalies. Microsoft Sentinel, which is a cloud-based SIEM solution built into Azure, has this feature.

Summary

Throughout this chapter, we discussed different types of techniques that attackers will use to compromise your privacy, various phases of attacks, and how information gathering is useful to conduct a successful attack by attackers. We listed the different types of tools that attackers will use to gather information, the important information that attackers will collect during another important phase, enumeration, and the various tools that can be useful. We also learned how identity exposure can be very critical for any organization. If devices are stolen or kept unattended, that can be a goldmine for attackers. On the other hand, if vulnerabilities are not available, attackers can create vulnerabilities using social engineering attacks.

In the next chapter, we will improve our knowledge of the different kinds of tools that attackers can use to compromise privacy.

Tools and Techniques That Attackers Use

In the previous chapter, we discussed various techniques that attackers use to compromise your privacy. This chapter will explain the different tools that attackers can use to compromise privacy by gathering information mainly from open intelligence and social media. There are many tools that attackers use to collect information about the target individuals and organizations. As we discussed in earlier chapters, publicly available sources contain a lot of information. Here, we will detail the tools and techniques employed by attackers to effectively collect information.

The following topics are discussed in this chapter:

- Maltego
- Google Advanced Search and dorks
- Uvrx Social Search
- Open Wi-Fi networks
- Phishing sites

Maltego

Maltego is a software developed by Paterva, a South African-based software company, to provide a framework to discover the data collected via **open source intelligence** (**OSINT**) and visualize it via easy-to-understand graphical representations. This is a very useful tool, as hackers or penetration testers usually try to collect as much information as possible before moving on to the next step. Maltego is treated as one of the best information gathering and data mining tools currently available. Users can query many types of data integrated with Shodan, VirusTotal, archive.org's Wayback Machine, TinEye, and MITRE ATT&CK. Maltego uses over 50 open intelligent data sources to provide information. It supports public data sources that you can connect to, as well as commercial data sources.

Maltego is one of the most frequently used tools by security testers, forensic investigators, investigative journalists, and researchers and comes pre-installed in many tiger boxes, including Kali, Parrot Security, and Predator. Maltego is a Java application that is available for Windows, Linux, and Mac in three editions: Community, Professional, and Enterprise. The Community edition is free and has only 5% of the functionalities of the Professional or Enterprise editions according to Paterva.

Maltego can gather information from dispersed data sources. Some versions of Maltego can view up to one million entities on a graph. It can collect information from connected data sources, automatically interlink them based on RegEx algorithms, autodetect entity types to enrich the graphical interface, and use entity weights to detect patterns that can be used to annotate the graph and export it to various formats that can be used for later reference.

Maltego as an information-gathering tool provides real-world links between many entity types. That can be very useful during information gathering as it details relationships that include the following:

- People
- Social networks or groups
- Organizations
- Companies
- Websites
- Internet infrastructure information such as domain names, DNS names, and IP addresses
- Phases
- Documents and files

Maltego reveals links between entities using OSINT. Transform hubs are sources that Maltego uses to acquire information.

Figure 5.1 - Maltego (Community edition)

Maltego uses a range of transform hubs to integrate with many OSINT data providers in its free and commercial editions. Depending on your usage requirements and choices, you can connect free editions of data sources or paid premium editions. As an example, both VirusTotal Public and VirusTotal Premium **Application Programming Interfaces (APIs)** for integration can be connected with Maltego.

Maltego can create graphs when you perform OSINT for a domain name or DNS system. There is a long list of options available in Maltego to be used as entities including organization, company, DNS name, domain, MX records, URL, website, location, hash, port, people, services, and more.

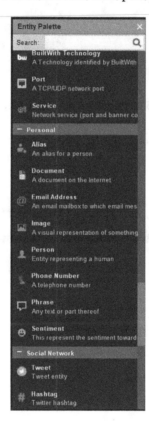

Figure 5.2 – Entity types in Maltego

Then, you can select any entity type you want to investigate, as shown in *Figure 5.2*. Let's say we are going to investigate a domain name. We can drag and drop the entity type to the graph window. If we need to investigate microsoft.com, we can type microsoft.com as the domain name and right-click to get transforms. Maltego has a built-in list of transforms, or we can add transforms using the Transform hub. When you execute the transform, Maltego connects through integrated APIs and collects OSINT information, which is then displayed in the graph window:

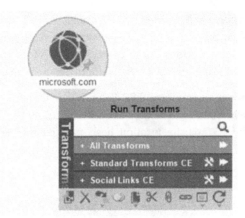

Figure 5.3 – Transform types in Maltego

When we click on the play button-style icon in front of the desired transform type, it will start transforming the information and display it in the graph window. *Figure 5.4* displays a visual representation of the information collected by Maltigo:

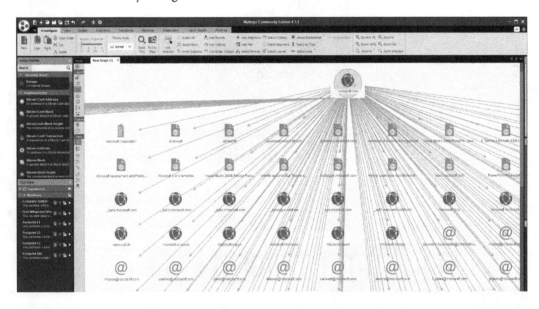

Figure 5.4 – A graphical representation of the information collected using OSINT in Maltego

In the same way, you can use a range of entity types, including email addresses, company names, URLs, and WHOIS data to transform the information of the target using Maltego. If you look at the available entity types and transformation hubs, you will understand how anything exposed to the internet matters. When it comes to information gathering, almost any kind of information exposed to the internet can be used by OSINT sources to collect information and understand the connection between different entity types. This has become very easy now, as Maltego provides visualizations and easy-to-understandable graphical forms, which allow security testers to easily interpret and co-relate entity types.

A Maltego-generated graph can be a gold mine for an attacker, as it provides interrelated information about entities. Especially when planning a social engineering attack or information gathering, this proves useful.

Google Advanced Search and dorks

Google is the most popular search engine, capable of searching through the searchable portion of the internet and providing accurate and most relevant results much faster than other search engines. Google frequently crawls through all the websites in the world and keeps cached copies of the web pages in Google Cache. Then, Google indexes all the web pages according to the keywords. The results we derive from a Google search are not something Google has searched on the go but have been searched and indexed even before you execute your own search. That's the reason Google suggests you use keywords whenever you start typing a query on Google search.

Dorks

There are ways that we can use the hidden capabilities of Google by employing special syntaxes, which are known as Google dorks, or as Google hacking. Attackers use Google dorks to gather information from the internet using Google Search capabilities.

As an example, let us check what Google has in the cache for www.aljazeera.com. The reason to use www.aljazeera.com to demonstrate this is that we need a frequently changing site to understand that a live copy can be different from the copy that Google has in its cache. We are going to search cache: www.aljazeera.com in the Google Search box:

Figure 5.5 – A copy of www.aljazeera.com with the snapshot information from Google Cache

There is a range of Google dorks that can be used to gather more sensitive information using Google Search. Let's try to understand some basic keywords that can be used in the Google dorks and then evaluate how to construct more advanced dorks to find out specific information. All Google dorks can be executed in the Google Search box in any popular browser. The following are some Google dorks:

- `inurl` – This command restricts the search to the URL of the web pages. For example, if you want to search for admin logins, you can use `inurl:adminlogin.php`. This will list out web pages with `admin login` in their URLs. All the web pages with `admin.php` will be listed.

- `intitle` – This command restricts the search to the titles of the web pages. If you need to view the list of items on a web page, you can use `intitle:"index of"`. This will list web pages with indexed items.

- `intext` – This command only searches the body text by ignoring the URL, titles, or links. If you want to find the text `cyber anonymity` irrespective of URLs and titles, you can use `intext:"Cyber anonymity"`.

- `site` – This command filters the search and restricts it to a single site. As an example, if an attacker wants to search for `satya nadella` only on `www.microsoft.com`, then the `site:www.microsoft.com "Satya Nadella"` command can be used. Google will not search anywhere other than the Microsoft website.

- `link` – This command returns a list of web pages linking to a specific website or web address. If you enter `link: www.microsoft.com`, Google will return a list of web pages that link to `www.microsoft.com`.

- `related` – This command finds pages related to other pages. If you search `related:google.com`, it will return a list of search engines.

- `cache` – If you use a frequently changing URL, you will see the page that Google returned can be different from the real page, as Google keeps a cached copy during crawling. `cache:www.microsoft.com` will show you a cached copy of the Microsoft home page that Google has on Google Cache servers.

Figure 5.6 – A Microsoft web page cached in a Google Cache server

- `filetype` – This command searches the filename or extension. You can search based on the filename and search for a specific file type. If you search `filetype:htm`, you can find a list of files with defined extensions.

You can also combine two search commands. `Application filetype:pdf` can search applications with the `.pdf` extension.

The following are some of the many examples of Google dorks that attackers use to find sensitive information:

- `inurl:adminlogin.php` – This will list out web pages with admin login in the URL.
- `inurl:"*admin | login" | inurl:.php | .asp` – This will list out php and asp pages with `admin login` in the URL.
- `intitle:"index of" inurl:ftp` – This will list out web pages with FTP file structures and indexes.
- `intitle:"Login" intext:"IP Surveillance for Your Life"` – This will search surveillance system logins.
- `intitle:"Blue Iris Remote View"` – This will list the "Blue Iris" cameras that are online, mostly without even having to provide credentials.
- `intitle:"web admin login" "Huawei Technologies"` – This shows admin login pages for Huawei routers and devices.
- `inurl:/multi.html intitle:webcam` – This shows the web applications with webcams.
- `"Username" "Password" "ZTE Corporation. All rights reserved."` – This shows ZTE appliance login screens.
- `allintext:"*.@gmail.com" OR "password" OR "username" filetype:xlsx` – This shows a list of web applications that has Gmail email usernames and passwords stored in `root` directories.
- `inurl:/wp-content/uploads/ ext:txt "username" AND "password" | "pwd" | "pw"` – This shows WordPress (wp) web applications with usernames and passwords stored in an `uploads` directory.

If you need to find more complex Google dorks, you can refer to the **Google Hacking Database (GHDB)**. This also contains community-developed dorks. Currently, GHDB is hosted at `https://www.exploit-db.com/google-hacking-database`. This source provides a long list of Google dorks that can be used for retrieving sensitive information using Google Search.

Figure 5.7 – The GHDB with a range of Google dorks

Google dorks are one of the most common ways of retrieving information from a website; even attackers can personalize Google to only perform an advanced search for a specific website by limiting the Google search scope for a single website using `site: www.sample.com`.

Google search engines crawl through the web pages to collect more specific information to provide an accurate result. If you want to limit search engines from crawling through sensitive directories of your website, you can use `"noindex"` placed on the meta tag and as an HTTP response header. To prevent web crawlers from indexing a page on your site, you can place the following `meta` tag into the `<head>` section of the page:

```
<meta name="robots" content="noindex">
```

Some developers think that they can keep sensitive directories out of Google Search by placing `robots.txt` with `allow` and `disallow` tags in the file for sensitive directories. However, this is used to avoid page overloading by search engine requests. If you type in any website and add `robots.txt`, you will find the directories that the crawler can still access. At the same time, the `robots.txt` file exposes the directory structure to attackers, which can then be used by the attackers for their benefit.

Let's try to test this by typing `www.microsoft.com/robots.txt` as a URL into the browser. The browser will display the contents of the `robots.txt` file as follows:

```
#Robots.txt file for www.microsoft.com

User-agent: *
Disallow: /en-us/windows/si/matrix.html
Disallow: /en-us/windows/si/matrix.html
Disallow: /*/security/search-results.aspx?
Disallow: /*/music/*/search/
Disallow: /*/store/buynow?
Disallow: /*/store/cart
Disallow: /*/search/
Disallow: /*/music/*/Search/
Disallow: /*/Search/
Disallow: /*/newsearch/
Disallow: *action=catalogsearch&
Disallow: /*/store/d/groove-music-pass/cfq7ttc0k5dq/0001
Allow: /*/store/*/search/
Allow: /*/store/*/layout/
Allow: /*/store/music/groove-music-pass/*
Allow: *action=catalogsearch&catalog_mode=grid&page=2$
Allow: *action=catalogsearch&catalog_mode=grid&page=3$
Allow: *action=catalogsearch&catalog_mode=grid&page=4$
Allow: *action=catalogsearch&catalog_mode=grid&page=5$
Allow: *action=catalogsearch&catalog_mode=grid&page=6$
Allow: *action=catalogsearch&catalog_mode=grid&page=7$
Allow: *action=catalogsearch&catalog_mode=grid&page=8$
Allow: *action=catalogsearch&catalog_mode=list&page=2$
Allow: *action=catalogsearch&catalog_mode=list&page=3$
Allow: *action=catalogsearch&catalog_mode=list&page=4$
Allow: *action=catalogsearch&catalog_mode=list&page=5$
Allow: *action=catalogsearch&catalog_mode=list&page=6$
Allow: *action=catalogsearch&catalog_mode=list&page=7$
Allow: *action=catalogsearch&catalog_mode=list&page=8$
Disallow: *action=accessorysearch&product=*&*
```

Figure 5.8 – robots.txt shows the directory structure of the target website

Even though `robots.txt` provides information to crawlers this way to prevent overloading, attackers use a `robots.txt` file to understand the directory structure of the web application. That gives attackers the chance to construct specific attacks such as directory traversal attacks. In some cases, the attackers even get access to password files on the server.

Google Advanced Search

Another powerful search built into Google is Advanced Search. You can access Google Advanced Search from `https://www.google.com/advanced_search` or by clicking on the cog wheel that appears in the top right-hand corner of Google once you perform any search. Once you access Google Advanced Search, it has many advanced features that a basic Google search does not possess. Using the advanced options, you can refine the search the way you want. For example, if you search for something from the United Kingdom, the search result is different from the same search someone performs from Australia, as Google refines the search to the location as well. If an attacker wanted to find out information that is relevant to a specific country scenario, the attacker could use the location in the search criteria as well as other options, including the following:

- All these words
- This exact word or phrase
- Any of these words or none of these words
- Numbers ranging from x to y
- Language

- The last update
- Region

Figure 5.9 – Google Advanced Search helps perform advanced searching

Google Advanced Search is quite useful in case you want to perform advanced searches and refine the results. For example, if you want to search for something that was updated last month or year, you can refine the search accordingly.

Reverse Image Search

Another great feature Google provides is Reverse Image Search. Similar to how you search for a word or phrase, you can use an image to search the entire searchable internet to find a similar image or the same image on different websites. You can perform an image search just by uploading the image to the tool or by providing the image URL. To access Google Reverse Image Search, visit `www.google.com/imghp`:

Figure 5.10 – Google Image Search helps to find similar images on the internet

These types of techniques are used by attackers to retrieve information from publicly available sources. These tools are immensely powerful, comprising the information available also known as OSINT. As explained in earlier chapters, information gathering is the first step in hacking. Attackers use these tools to uncover as much information as possible, which will make their next steps easier.

Uvrx Social Search

There are huge amounts of data stored on the internet, which is searchable by general search engines such as Google and Bing. The term **searchable internet** was purposely used in many chapters, as the internet possesses huge space besides a searchable internet, which is known as the **dark web**. This is not searchable using common search engines. *Chapter 9, Avoiding Behavior Tracking Applications and Browsers*, will provide more information about the dark web.

Many online file storage systems host large amounts of data separate from the dark web or the searchable internet. Many users make use of free or commercial versions of online storage to store and share files with themselves or between communities. These storage systems may also contain important and useful data.

Uvrx is a very comprehensive search engine initially designed to search online file storage. Currently, Uvrx supports three types of comprehensive searches:

- File Search
- Social Search
- Health Search

Uvrx File Search

Uvrx File Search is a comprehensive online file storage search engine that can be used to search within many popular online file storage services. Uvrx provides individual search engine capabilities for the following online storage services:

- `badongo.com`

- `mediafire.com`

- `zshare.net`

- `4shared.com`

- `taringa.net`

Figure 5.11 – The Uvrx search engine can search a range of online storage

Currently, the Uvrx search engine supports multiple languages, including English, Chinese, Japanese, French, German, Spanish, Portuguese, and Russian.

Apart from the individual search engines for the aforementioned online storage services, Uvrx also provides a *search all* feature, which searches across a range of other online storage providers, such as FileFactory, DepositFiles, EasyShare, sharedzilla, GigaSize, DivShare, Sendspace, YouSendIt, Badongo, MediaFire, zShare, 4shared, Letitbit, drop, FileSurf, Hotshare, USAupload, SaveFile, Bigupload, upfiles, HyperFileShare, Zippyshare, uploaded.to, uploading, sharebee, Rapidspread, Taringa, and more with the search query. Uvrx provides a free service to search across many online storage providers conveniently, rather than having to search individually.

Uvrx Social Search

Uvrx Social Search is a free and comprehensive search engine enhanced by Google that supports searching on multiple social networks. This is a convenient way to search social networks such as the following:

- Facebook
- Twitter
- Myspace
- LinkedIn
- Plaxo
- Instagram
- Tumblr
- LiveJournal
- Flickr

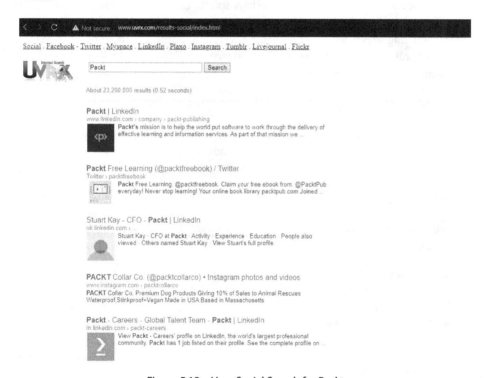

Figure 5.12 – Uvrx Social Search for Packt

Even though Facebook blocks user information that is marked as private from being accessed, crawled, and indexed by search engines, not all the information on Facebook is marked as private. This can be due to a lack of understanding on the part of the users about privacy, whether mistaken or intentional. Therefore, all information on Facebook including pages, posts, and updates that are set as public can be crawled. For example, currently, Google has over 2 billion Facebook pages in the Google index.

This can be used by various interested individuals, including attackers, who can find out juicy information on Facebook, or users who do not understand how to minimize cyber exposure by using the privacy settings on Facebook.

Uvrx is a great resource for you to understand your status in terms of privacy on social media. If the privacy settings are not properly configured on your social media profile, you can see how much information that you think is personal is actually visible to the public.

You can find out by yourself about this level of exposure by using the Uvrx Social Search engine.

Uvrx Health Search

Many websites on the internet are created for advertising purposes. For example, if you need to find information about software, many sites provide reviews and suggestions. These suggestions or reviews may not always be realistic and trustworthy, as many sites are made-for-advertising. The objective of made-for-advertising sites is to attract users for advertising purposes. These sites are funded by vendors to provide reviews on their products and advertising, depending on the number of page views. We cannot expect neutral reviews from these sites, as they are always biased in favor of the companies that provide the funding. Moreover, made-for-advertising sites flood the internet, as they provide passive income generation for many individuals and companies.

When it comes to health-related concerns, this is more critical. For example, if you are searching for information about a specific illness or a drug, made-for-advertising sites provide biased information that is not trustworthy or reliable a lot of the time.

This makes it very difficult for individuals to find trustworthy, reliable, and unbiased information on the internet. Uvrx Health Search tries to solve this issue by maintaining a list of trustworthy and unbiased websites maintained by governments, trusted organizations, educational institutions, and research organizations.

As an example, let's imagine an individual is experiencing numbness in their toe. If the person searches this on the internet to find information, websites will produce many different suggestions. Most of these websites are backed by pharmaceutical companies to promote their products. They can even suggest different medicines and ideas that are often confusing. Some websites may say it is a malnutrition issue, some a head injury, or even diabetes. But most of these websites are biased, as they are funded by companies or these companies advertise on them. In these situations, it's difficult to understand which website will give you an honest opinion. Uvrx Health Search can be a lifesaver in this context.

Open Wi-Fi networks

With today's complex lifestyles, connectivity has become an essential requirement. Before the 21st century, food, water, and shelter were treated as essentials. Later, electricity became an essential requirement. Now, in today's world, besides food, water, shelter, and electricity, connectivity has become an essential requirement. This is the reason hotels and restaurants, for example, highlight their connectivity status.

If you search for any hotel for your holiday on `booking.com` or `tripadvisor.com`, all the hotels show whether they provide internet access.

Most airports, restaurants, libraries, malls, and open areas provide Wi-Fi connectivity, as most individuals cannot stay disconnected. Even long-distance bus services and flights provide Wi-Fi access today as a complimentary service to attract customers.

In *Chapter 1, Understanding Sensitive Information,* we discussed the layers that are required for communication. If you need to maintain secure communication, we need to protect all the layers. If any of the layers are not secured to an adequate level, communication can be compromised.

What is open Wi-Fi?

Many studies conducted around the world, including those by security research companies such as Kaspersky labs, have revealed that 25% of the world's public Wi-Fi hotspots don't use any type of encryption to protect their communication. These hotspots are generally referred to as **open Wi-Fi** connectivity. The public comprehends this term as *open for anyone* or a *password not being required to connect to the internet.* Due to this reason, they are very happy to find these open Wi-Fi networks and connect their devices to them. But the bitter reality is that open Wi-Fi networks do not have any wireless security protocol configured. This allows anyone to connect to the internet using the open Wi-Fi network and an attacker connected to the same network can sniff all the communication if the communication protocol used by the user is unencrypted. When you connect your device to the internet, there are many applications installed in the device to start communication using internet connectivity. These applications may or may not use security protocols. For example, if you configure an email client app installed on your device to use **Simple Mail Transfer Protocol (SMTP)**, the communication is unencrypted unless **Transport Layer Security (TLS)** is enabled.

This can happen in coffee shops and other open restaurants such as McDonald's. One of the most frequent questions customers ask is "what's the password for the Wi-Fi?" It is quite common to see restaurants display Wi-Fi passwords publicly or print them on their invoices. Most restaurants have common passwords or open Wi-Fi hotspots.

Most importantly, every user must understand the importance of the sensitive data they carry within their devices. When connected to the internet, applications can continue communication as usual, even though users think otherwise. Users often think that only the applications that they use are currently connected to the internet, whereas the reality is that any of the applications installed on the device can communicate when connectivity is established.

Risks involved with open Wi-Fi

When connected to open Wi-Fi for communication, everyone should know the risks and threats, as there could be situations where we might need to connect to open Wi-Fi:

- **Insecure connections** – As open Wi-Fi does not configure with secure protocols, attackers can sniff and monitor your communication. If you connect to a website that provides secure connectivity using **Hypertext Transfer Protocol Secure (HTTPS)**, the connectivity from your browser to the server is encrypted by a key. The browser-generated key is shared using the digital certificate installed on the server:

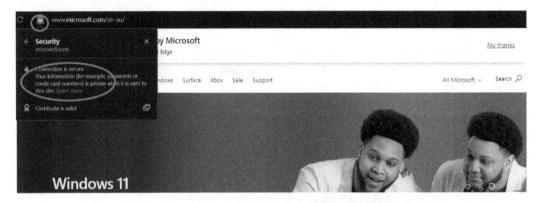

Figure 5.13 – A HTTPS secure connection from a browser to a web server

When the digital certificate is obtained and installed on the web server, it provides a secure connection from the browser to the server. This can be seen on the address bar with the padlock sign (*Figure 5.13*). When you click on the padlock sign, you can see the connectivity status. If you click on the certificate, the browser will show you the certificate information.

When the HTTPS connection is used on open Wi-Fi, the connection between the browser to the server is encrypted. This reduces the chances of the attacker monitoring your communication, including your credentials.

Not all the websites in the world are encrypted. Unencrypted websites can be detected by browsers and be shown as **not secure**. This means that when you send and receive information from this website, data sent back and forth is not encrypted. If an attacker is sniffing, they can see and intercept the data, which is in the form of plain text:

Figure 5.14 – A HTTP insecure connection from a browser to a web server

- **Personal information can be theft** – This is a very common threat in open Wi-Fi, where attackers collect personal and sensitive information during communication, including the following:

  - Login credentials

  - Personal data

  - Private images

  - Financial information

  - Confidential information

  This can be done simply by sniffing. If you use an unencrypted protocol for communication, attackers can simply sniff your information using a sniffer such as **Wireshark**. Wireshark is a free and open source protocol analyzer and sniffer that supports various operating systems. Once an attacker starts sniffing, all plaintext information sent through the communication can be sniffed.

  For example, let's use Wireshark to sniff unencrypted traffic. We will use previously learned techniques to find out a website with the following requirements:

  - A website that only uses HTTP and not HTTPS

  - We need to find the login page

  This can be done using Google dorks. Let's form a Google dork to find a website with these requirements. Here is the Google dork that we created; this may be different from your Google dork, as the same thing can be performed in different ways:

```
Inurl:http -inurl:https inurl:adminlogin.php
```

Once the preceding Google dork is entered on Google Search, Google comes up with a list of web pages matching our requirements. Then, we can select one of the web pages and notice that it doesn't have HTTPS, and it's a login page:

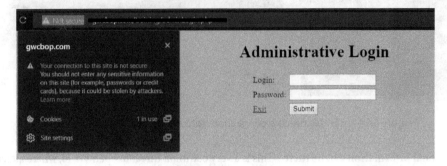

Figure 5.15 – A HTTP insecure connection from a browser to a web server

Now, we will run Wireshark by selecting the network interface and start monitoring the traffic. Then we will enter a username and password (yes, we don't know the credentials, so we will enter random credentials). Let's check on Wireshark. You can see a lot of information on Wireshark; to search for the exact information, we will use search:

- Select **Edit | Find Packet**.

- Select **String** and type POST.

Then, Wireshark will highlight the answer. The reason to search for POST is that it is one of the two methods that web applications can use to communicate with the browser. Typically, the POST method is used by application developers to send credentials back to the web server.

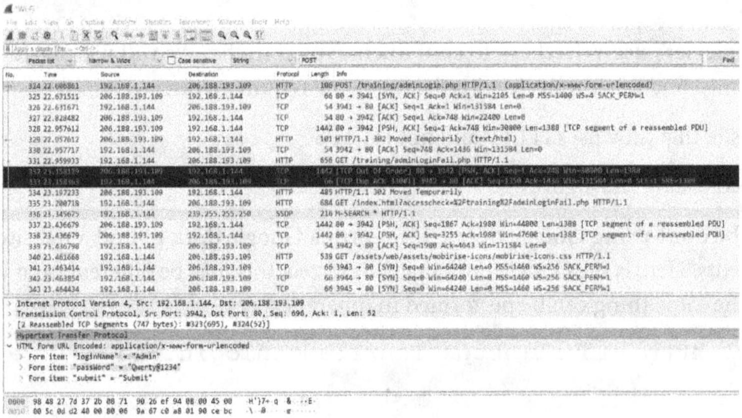

Figure 5.16 – Wireshark collects the credentials from an insecure
communication between a browser and a web server

This shows how easily a protocol sniffer can collect sensitive information from unencrypted communication. When you are connected to open Wi-Fi, attackers can easily collect your sensitive information, as shown here.

- **Man-in-the-middle attacks (MITM attacks)** – Attackers can create a **rogue access point** (using the same SSID, attackers can create an unauthorized hotspot that they can use to convert their own device into a hotspot) or an **evil twin** (a cloned access point with the same SSID). If the open Wi-Fi system is present, it's easy for attackers to launch MITM attacks.

- **Organization-targeted attacks** – To access confidential organizational information, attackers can use open Wi-Fi. Some organizations use secure encrypted Wi-Fi networks within the corporate network and create open Wi-Fi for the guest users without distinguishing network segments. This configuration provides attackers with the possibility to access corporate networks over open Wi-Fi. Another possibility is that internal employees can often also use open Wi-Fi to bypass any filtering enabled for corporate users. If the attackers are monitoring the guest network, they can often collect confidential information from negligent users.

- **DNS poisoning** – Once attackers are connected to the open Wi-Fi, they can redirect user connections to the attacker's desired web applications, including phishing sites, using DNS poisoning. For example, let's say one of the users wanted to access `www.facebook.com`. When a user types `www.facebook.com` into the browser, the browser requests the DNS client of the operating system to resolve the IP of `www.facebook.com`. Then, the DNS resolver sends this request to the DNS servers if the required IP is not present on the local DNS cache. An attacker can listen to the DNS query sent to the DNS server and replace the result with the attacker's desired IP. Then, the DNS resolver will receive a fake result, which is known as DNS poisoning. The browser will then try to connect to the fake phishing site using the attacker-replaced IP.

- **Session hijacking and malware distribution** – When a user connects to a website and a session is established after authentication, attackers can hijack the session and connect to the web application using a stolen session ID. Whenever a user enters valid credentials to a web application, the user will be authenticated by the web application. Typically, this creates a session for the user. Once the authentication is complete, the session ID will be used to control the user session. For example, some web applications such as banking applications are designed to terminate the session if the session is inactive. Using a packet sniffer or protocol analyzer such as Wireshark, attackers can collect the session information and rewrite session information to their browser to reestablish the connection without even using the credentials of the genuine user.

In the same way, attackers can redirect users to fake sites with malware and distribute malware among the users. Since users have entered the correct address in the browser, they think the website they have accessed is genuine, but in reality, it could be a phishing website created by the attacker to lure users and distribute malware.

How to minimize the risks with open Wi-Fi

Open Wi-Fi networks have various risks that we have discussed in this chapter. But there are unavoidable situations where we will still have to use open Wi-Fi systems, even though we know that there are multiple risks involved. Let's look at the ways you can minimize the risks involved when using open Wi-Fi. For example, when you travel overseas and are in transit, you may need to send an urgent email. Maybe you need to transfer some funds from your bank account urgently and the only option that you have is through airport Wi-Fi. Some airports provide open Wi-Fi for their passengers while some airports, such as Changi airport in Singapore, provide authenticated **one-time passwords (OTPs)** to their passengers after scanning their passports:

- **Use Virtual Private Networks (VPNs)** – VPNs are one of the secure ways to access the internet and its services more securely, as it creates encrypted communication. There are open and commercial VPN solutions. Most VPN solutions support mobile devices and some even support multiple devices with a single license. When connected to open Wi-Fi, you then need to connect to a VPN. This will create an encrypted connection from your device using a VPN client to the server. Some VPN providers even let you use your desired server location:

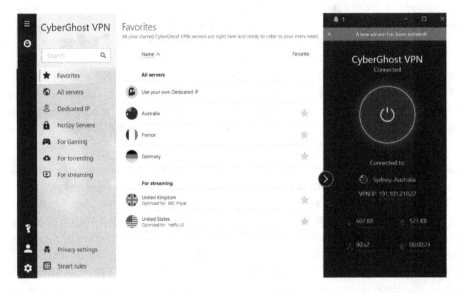

Figure 5.17 – The Cyberghost VPN provides secure communication even if connected to open Wi-Fi

When you start your communication over VPN, communication is encrypted from the device to the server. Chances for attackers to sniff the communication will be drastically reduced. When you are connected to open Wi-Fi, a VPN such as the Cyberghost built-in module will automatically start protecting the communication:

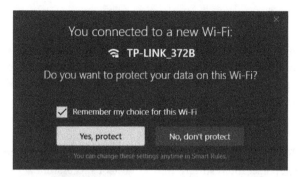

Figure 5.18 – The Cyberghost VPN Wi-Fi module secures the communication

Using a trustworthy VPN service provider is important. There is a range of VPN tools that can be used to encrypt communication when connected through open Wi-Fi. You can select VPN services such as the following:

* NordVPN

* OpenVPN

* ProtonVPN

* NetMotion

* ExpressVPN

* **Don't send private or sensitive information over open Wi-Fi** – As a rule of thumb, we should refrain from sharing sensitive and private information over open Wi-Fi. We can reduce the risk by connecting over VPN connections, but always try not to send **personally identifiable information** (**PII**) over open Wi-Fi.

* **Multi-factor authentication** (**MFA**) – Enabling MFA whenever possible is another way of reducing the risk. If you enable MFA, and even then, attackers collect your credentials, they will not be able to access your financial information or sensitive accounts, as they need another level of authentication. Most banks and social networking tools support MFA.

* **Choose a cellular network over open Wi-Fi** – In today's world, most users are used to connecting to hotspots, including open Wi-Fi, whenever available, even if they have cellular networks. Mostly, this happens due to the speed and capacity limitations of cellular networks. But when it comes to security, your own cellular connection is more trustworthy than unknown open Wi-Fi or hotspots, as it's easier to manipulate Wi-Fi than cellular networks. If you need secure connectivity, always use cellular networks or mobile hotspots instead of open Wi-Fi.

Phishing sites

Phishing is another technique that attackers commonly employ to compromise sensitive information. Phishing is a process of attempting to acquire sensitive information by masquerading as a trustworthy entity. Phishing can be used as an open attack or targeting attack. When a targeted attack is conducted on an organization or a user, then the attack is known as spear phishing. We have discussed various ways that attackers use phishing techniques and tools in the *Social engineering* section of *Chapter 4, Techniques that Attackers Use to Compromise Privacy*. The first-ever known phishing attack was reported in 1996, but a paper published in 1987 by the HP user group described a phishing technique 10 years before the real attack took place. Interestingly, in over 90% of successful data breaches, phishing is involved in some way according to the well-known security research company, KnowBe4. Phishing has over three and a half decades of history and has been constantly evolving in this time.

Newer phishing attacks are more sophisticated and innovative. Some developments that phishing attacks have seen include the following:

- Links sent by email addresses with fake Google search results, which redirect the user to attacker-controlled websites, mostly malware-laden

- Phishing company-controlled cloud logins, including Microsoft 365, that look real, so users will not think twice before entering their credentials

- Redirecting legitimate users to malware-laden, attacker-controlled sites using these aforementioned techniques and infecting the endpoint devices with malware

Summary

This chapter explained the different tools that attackers use to compromise privacy, including Maltego, Google Advanced Search, Google dorks, the Uvrx search tools, open Wi-Fi networks, and phishing attacks. This chapter also provided knowledge on how attackers use Google Search to collect sensitive data, how to search for anyone across social media, the risk of open Wi-Fi networks, and how to protect yourself from phishing.

In the next chapter, we will explain the types of data that competitors are interested in collecting from companies and individuals. We will be learning about various artifacts that attackers collect from you, the artifacts that competitors are interested in collecting from individuals and from companies, the ways that these attackers can access your networks, and how attackers can compromise browsers.

6

Artifacts that Attackers Can Collect from You

Traditionally, IT security and cyber security professionals concentrated on protecting the perimeter. The perimeter is the area in our infrastructure that separates the internal infrastructure from the external world. This practice is commonly known as **castle security**. It was used by conventional security systems to try to protect information from external perpetrators. Most systems were closed systems and the only connectivity to the external world was the internet; this practice was quite acceptable for traditional systems. However, today's complex requirements and the introduction of cloud systems have made this practice worthless. Also, insiders play an active role in many attacks, meaning security professionals are forced to find a better approach. Since attackers must also access enterprise systems through the perimeter, security professionals can collect indicators of compromise from perimeter devices.

Typically, artifacts are what get left behind after an activity. We can treat them as footprints of the end user or an attacker. This chapter focus on the artifacts that can be collected from you by attackers and other interested parties.

In this chapter, we will explore the following topics:

- Artifacts that attackers can collect from you
- Artifacts that companies would like to collect
- Devices that can be compromised
- Ways that an attacker can access your networks
- Compromising browsers

Artifacts that attackers can collect from you

Attackers can collect artifacts from you in many ways. Traditionally, artifacts were used by cyber-forensic professionals to trace back perpetrators after an attack. In today's modern world, artifacts are quite useful to collect information about targeted users or companies to understand their behavior patterns, buying patterns from online stores, interests, and other exciting information. Artifacts can be collected by attackers on any device including desktops, laptops, or mobile devices. They can be collected from deep inside the operating system, memory, temporary files, and various locations inside the filesystem. Artifacts can provide significant information about the activities performed by cyber attackers, as well as users. This is the reason why artifacts are the main source of evidence for cyber-forensic professionals when analyzing an attack. For forensic professionals, artifacts provide information related to unauthorized access, tools that are installed or used by the attackers, attacker locations, and other relevant information, including attackers' IP addresses.

On the other hand, if attackers collect artifacts from users, they also can collect information about user activities.

Attackers use different ways to collect artifacts from target systems. Artifacts from the targeted systems that are commonly collected by attackers include the following:

- **Cookies**: Attackers use cookies to understand the locations you frequently checked into, login information such as session IDs, and products you have clicked on. Cookies are files that are very small in size, created by websites that you have visited and stored within your browser. These cookies allow websites to monitor your behavior to streamline your searches and provide you with a rich experience.

- **Files that are created by operating systems**: **Security Accounts Manager** (**SAM**) on the Windows operating system and `/etc/shadow` on the Linux operating system.

- Files that contain browser preferences, passwords, and history.

- Files that contain sensitive and personal information.

Let's look at the ways attackers can steal artifacts from your computer:

- **Malware attacks**: Malicious software, commonly known as malware, is the most commonly used way attackers steal artifacts from your computers. Hackers create malware to enter systems and perform other malicious activities. It's like throwing a net to catch fish. Whoever is infected by the malware provides attackers access to the infected system, allowing them to steal information and collect artifacts. Malware uses different entry points to get itself planted on targeted systems. It can be a vulnerability that is exploited by the malware, a link sent to a user that they clicked on to unwittingly install the malware, an attachment sent to the user's mailbox that is opened by them, or removable media that contains malware that is attached to the system. Malware is the common name used to refer to a large variety of programs of a similar nature. Let's look at some of the different categories of malware:

- **Viruses**: Viruses can replicate themselves and infect other programs. Viruses need host files to exist and are often attached to another executable. Nowadays, even script kiddies (attackers with little or no knowledge of advanced technologies that use existing tools to launch attacks) can create viruses using construction kits.

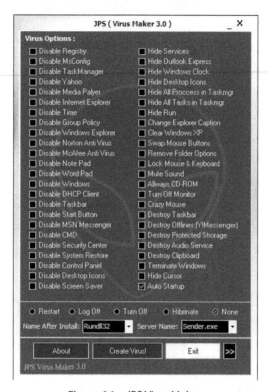

Figure 6.1 – JPS Virus Maker

JPS Virus Maker is a virus-constructing kit with which you can create your own virus. You can disable any program or process of a Windows system using this construction kit, including turning off the firewall or Windows Defender. As you can see in the preceding screenshot, an attacker can use this type of construction kit to create a virus. If they are an elite hacker, the attacker can write their own code to do the same.

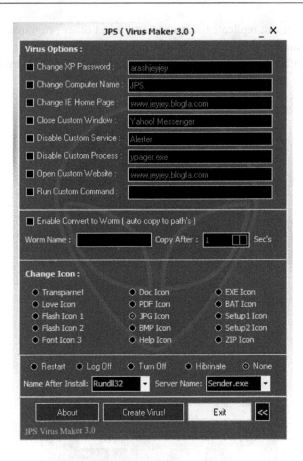

Figure 6.2 – More features of JPS Virus Maker

JPS Virus Maker has even more advanced features, such as changing a computer's name and running custom commands once executed. You can even convert the output to a worm, backdoor virus, or logical bomb. Once everything is configured using the GUI, an attacker can choose any icon they want, create the virus, and start sending it to the targets.

- **Worms**: Worms don't need host files as they are self-contained programs. Often, worms propagate through networks to infect other systems. Mostly, worms use emails to attach themselves and send copies to email recipients. As well as virus construction kits, there are worm construction kits available – for example, Internet Worm Maker Thing.

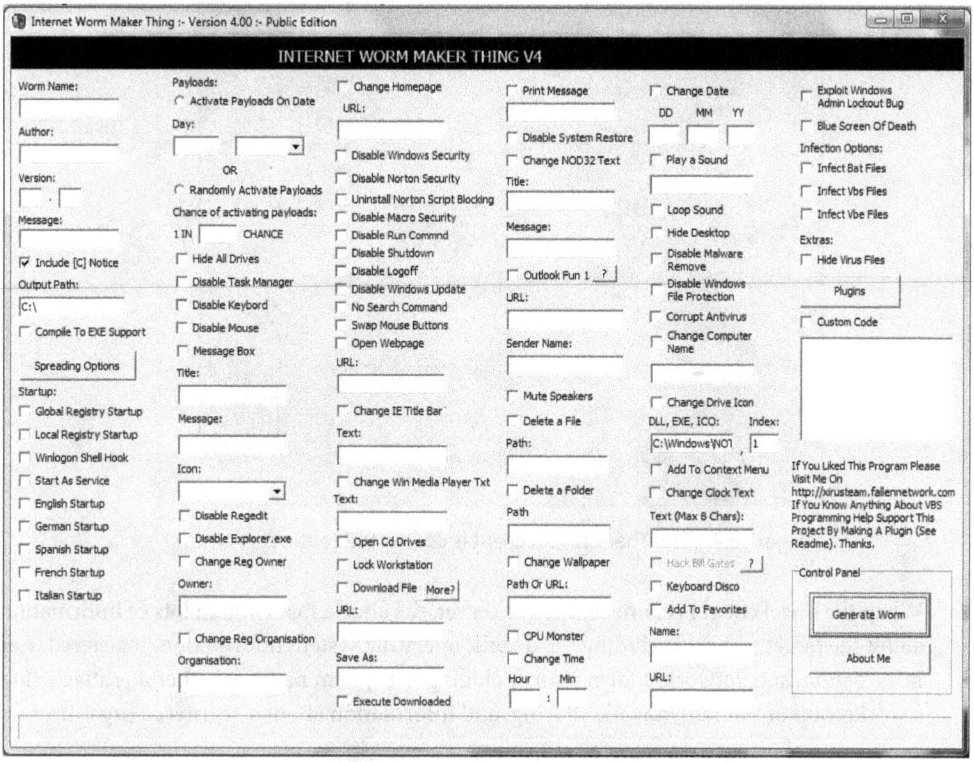

Figure 6.3 – Internet Worm Maker Thing to create worms

Internet Worm Maker Thing is a GUI tool that generates malicious code or automated script the way you configure it. You can define the criteria using the given options, including actions that you want the worm to perform. You can change the startup of the program, change the home page of the browsers, drive icons and names, change wallpapers, and many other options, as you can see in the preceding screenshot. Once you create the worm, it will be in Visual Basic Script format or .vbs. You can convert .vbs to .exe if you want to.

- **Trojans**: Trojans are malicious code hiding inside a legitimate program and performing malicious activities. Trojans don't replicate themselves like viruses and mostly create backdoors for attackers to enter and connect to a target system. This provides them with access to the system to collect artifacts from it. When it comes to collecting artifacts from systems, Trojans are the most commonly used malware by attackers.

As with viruses and worms, there are many Trojan construction kits available. Theef is one of the Trojans that attackers use to create a backdoor connection to a target machine. There are two parts to this Trojan, a server and a client. The attacker executes the Trojan server on the target system, and once the Trojan server is executed, the Theef client can be used to connect the target system to the server to control it, as shown in *Figure 6.4*.

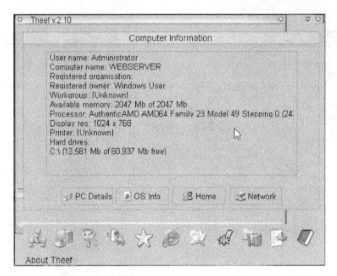

Figure 6.4 – The Theef Trojan client is connected to the server

When the Theef client is connected to the server, the attacker can obtain lots of information about the target system, including PC details, operating system information, registered user information, and network information, including the system name, whether it is attached to a workgroup or domain, any file sharing, and information about a registered organization.

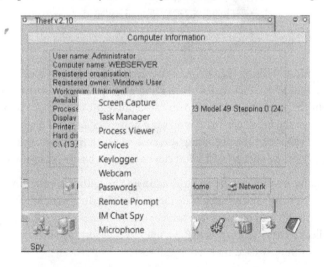

Figure 6.5 – The Theef Trojan can spy on the target

Once connected, the Theef Trojan can capture the screens of a target system, view and control the Task Manager and the Process Viewer, work as a keylogger, obtain passwords saved on the system, and remotely access Command Prompt, internet messenger chats, and the microphone.

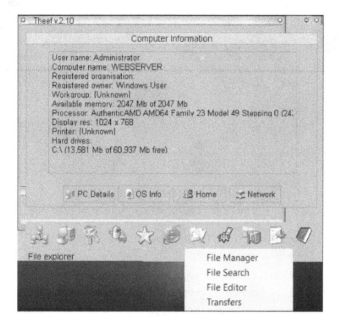

Figure 6.6 – The Theef Trojan can access the filesystem to collect artifacts from the target

Importantly, the Theef Trojan can access the filesystem of a target system and collect artifacts. This is a very important feature that the Theef Trojan has; it can search for important files of the target system, access the files, remotely execute them, open them, and transfer them to the attacker's system. This is a great way of collecting artifacts from the target system.

- **Spyware**: Spyware is another common malicious tool that attackers often use to collect sensitive information and artifacts from a target. Spyware is used in many devices, including mobile devices, to collect artifacts. There is a lot of spyware that attackers use to collect information from target systems, including commercial spyware. SpyAgent is one of the most commonly used commercial spyware software, which can spy on a target system using an open or stealth method.

Figure 6.7 – SpyAgent can collect artifacts and information from the target and send them to the attacker

As you can see from the preceding screenshot, Spytech SpyAgent can monitor a target system and extract information from the target to the attacker. This information includes keystrokes and mouse movements, windows accessed, program usage, screenshots, event timelines, access to files and folders, email and internet activities, and chats on social media. Attackers can also set advanced configurations, as shown in the following screenshot:

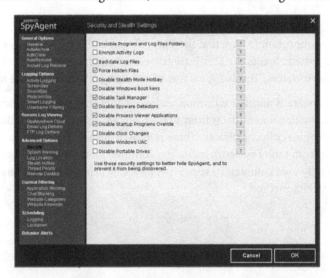

Figure 6.8 – SpyAgent advanced configuration that can be configured by the attacker

If attackers want, they can run the program in stealth mode so that a victim will not have any idea that they are being spied on and monitored. This also includes remote log delivery and viewing; attackers can configure SpyAgent to send logs to a specific email address, **File Transfer Protocol** (**FTP**) location, or spyware cloud so that they can monitor them remotely. Attackers can block applications, chats, and specific web categories from a target system. Attackers can even use this tool to compromise credentials through keylogging and screenshots. Attackers will receive credentials and sensitive information by email or FTP accounts.

SpyAgent introduced a feature called smart logging, which can log only special events, such as a user entering credentials, web addresses, and keywords. Once the attacker configures the schedule, SpyAgent sends emails at a scheduled time to them.

- **Ransomware**: This type of malware will encrypt important files when infected and make a ransom demand to release a key to unencrypt files. This is a very common type of attack these days, as attackers are more interested in quick financial gain from compromised systems.

- **Compromised credentials**: Usernames and passwords are the more common credentials used on many systems. Attackers are interested in collecting artifacts from systems using compromised credentials. Credentials can be lost, stolen, or exposed by other attacks, which attackers can use to access target systems. If the users are using weak credentials, guessable credentials, or compromised credentials, attackers can exploit the weak point. This includes system credentials, online web application credentials, and online services such as FTP credentials.

- **Phishing and spear-phishing attacks**: As discussed in previous chapters, attackers use sophisticated phishing attacks against individual users and companies mainly to compromise credentials.

- **Session hijacking attacks**: An attacker can take over a session established in client and server communication. In some attacks, the attacker steals the session ID and recreates it from their system, taking over the session without any usernames or passwords.

- **Application-level attacks**: Cross-site scripting and SQL injection are common application layer attacks that attackers use, especially crafter scripts or queries, to access sensitive information stored in a backend database connected to an application. If successful, the attacker even gets to control the database and the data stored in it.

Artifacts that competitors would like to collect

As discussed in the previous section, attackers are interested in collecting artifacts from you. These artifacts contain valuable information about you. These artifacts are used to find more information about targets. Some artifacts even contain the buying and surfing patterns of a user. For example, there are artifacts that keep information about most frequently visited websites, items of interest, and items purchased over the internet. For this reason, competitors are also interested in collecting artifacts from individuals and competitor companies. Mainly, they use cookies, known as third-party cookies, to collect information about competitors and individuals. If you understand how cookies work, then you can block and prevent disclosing your information as an individual or a company.

These tracking cookies can be categorized as direct web tracking cookies and third-party tracking cookies.

Direct web tracking cookies

Direct web tracking cookies are used to track visitor information and browsing information on a website directly. This helps to analyze visitors who return to the same website and their access behavior on the website. The cookies also track different pages that are visited on the same website. This allows website owners to provide a rich experience to web surfers when they visit the same website.

Third-party tracking cookies

Third-party cookies are not created by a website directly; typically, third-party cookies are created by an external server or service through embedded code on the original website. Third-party cookies are created by mass advertisers, by data analytics services, and through advertisements and plugins.

What can tracking cookies do?

As tracking cookies are typically used for advertising purposes, they collect data related to users' web-browsing behavior. Tracking cookies can collect information, including the websites you visited, web pages accessed on each website, items that you were interested in, files you download, images that you click on, location information, and device-specific information. Companies and vendors can send you personalized advertisements using the data collected by tracking cookies. Typically, cookies are not as harmful as viruses or worms, and they cannot generally make your computer open to infection by other malware. This means they cannot tamper with the computing process, or the way a computer usually works. However, one danger is that attackers can design viruses and worms that look like innocent cookies.

Another concern is that **internet service providers** (**ISPs**) and large marketing companies can create extraordinary cookies that have additional capabilities compared to normal cookies. These cookies can even recreate themselves after being deleted. These cookies are generally inserted into HTTP headers by ISPs. They use these cookies to collect information about browsing patterns and activities. The main concern is that most users have no idea that these cookies are tracking all their online behavior and passing them to specific companies.

How to handle cookies

Every time you visit a website, the website will create at least one cookie on your browser. This cookie is known as a direct cookie. A cookie created by the website you visited will then remember all your basic activities on that website, including how long you stayed on a web page and the number of pages visited on the website. A direct cookie will not record information about any other websites that you visited. Information collected by the direct cookie is limited only to the website you visited.

Third-party cookies are different from direct cookies. If a third-party cookie is stored on your browser, it allows advertising companies, social media companies, ISPs, other websites, and many other service providers to track your online activities and browsing behaviors. Third-party cookies can be stored on your browser by many websites. This is a direct threat to your online privacy. In other words, if you are concerned about your online privacy, you must block third-party cookies.

We must block third-party cookies from our browser, so let's see how. The settings can differ from browser to browser. However, all the popular browsers use similar options to block cookies, although the navigation and terminology can be a little different.

Disabling third-party cookies on Google Chrome

The default setting on Google Chrome is **Block third-party cookies in Incognito** According to Chrome, while in Incognito, *"Sites can't use your cookies to see your browsing activity across different sites, for example, to personalize ads. Features on some sites may not work."* (Refer to the following screenshot for reference.)

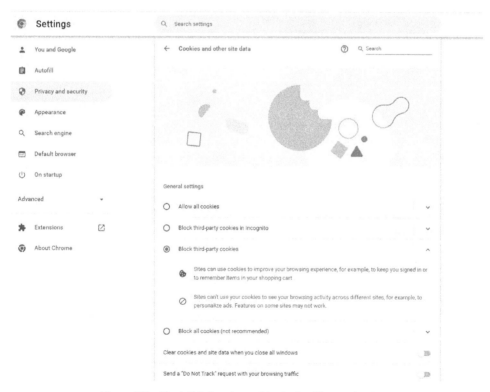

Figure 6.9 – Block third-party cookies in the Chrome browser

As a temporary solution for this, we can use the private browsing capabilities of the browser. Google Chrome has Incognito, Microsoft Edge has InPrivate browsing, and Mozilla Firefox provides Private Browsing. When you use a private browsing option to surf the internet, websites cannot use cookies to collect information of your browsing activities and personalized advertisements won't work. However, even if you use private browsing, websites can still collect information, such as the IP addresses that you use to access the websites and the specification of the device that you are using.

Disabling third-party cookies on Microsoft Edge

You need to click the three dots in the top-right corner of Microsoft Edge and select **Settings**, then select **Cookies and site permissions** on the left menu, and click on the **Cookies and data stored** option. Third-party cookies are allowed by default. If you want to disable third-party cookies, you need to enable the switch next to the **Block third-party cookies** option.

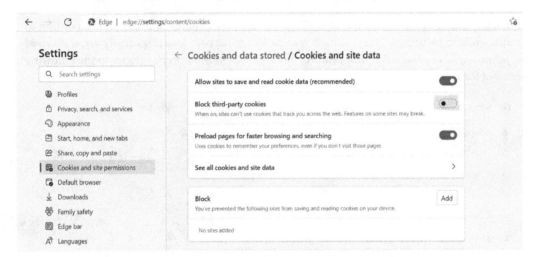

Figure 6.10 – Blocking third-party cookies in the Microsoft Edge browser

Disabling third-party cookies on Mozilla Firefox

Similar to Microsoft Edge, you need to click on the three lines in the Firefox browser's top-right corner, and then select the **Preferences** menu. You need to select **Privacy & Security**, and under the **Enhanced Tracking Protection** option, select **Custom**. Under the **Cookies** dropdown, select **All third-party cookies**. This will block all third-party cookies when you are using the Mozilla Firefox browser.

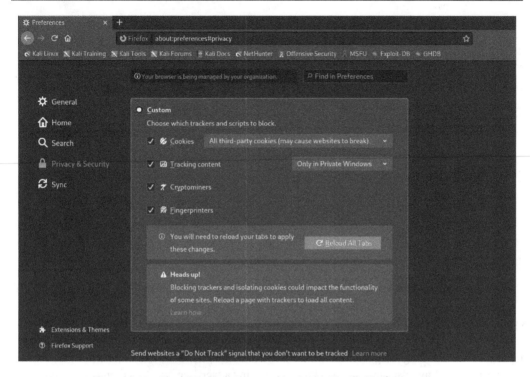

Figure 6.11 – Blocking third-party cookies in the Mozilla Firefox browser

Another option is to use secure browsers such as Brave, which was created by Brendan Eich, the founder of JavaScript, which provides total control over your own security and privacy. The Brave browser doesn't collect user information by default, and it is privacy-focused. Brave is built on Chromium, which was a Google-initiated, open source project from which many other browsers, such as Vivaldi, Microsoft Edge, Opera, and Google Chrome, were created.

The Brave browser

The Brave browser was created from a Google-led, open source Chromium project. The Brave open source browser is developed by Brave Software. Brave was developed with privacy and security in mind. Its browser is faster than other popular browsers. When considering all the security aspects of popular web browsers, Brave is the best for all aspects of security as a browser. As an open source browser, Brave has many useful built-in security features, including ad blocking, script blocking, auto upgrades to HTTPS, blocks against third-party cookies and other storage, and blocks against user fingerprinting. The main advantage of Brave is that all its privacy features are configured by default, whereas with other browsers, users usually need to configure privacy features.

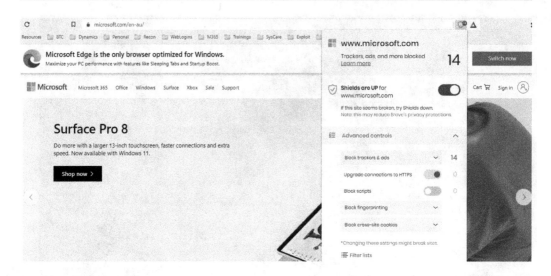

Figure 6.12 – Block trackers and ads in the Brave browser

As you can see in the preceding screenshot, Brave blocks all trackers and ads, and upgrades connections to HTTPS. If prefer, you can block scripts too. The preceding screenshot doesn't show blocking scripts, as when blocking scripts is enabled, most of the trackers are also be removed. Apart from these capabilities, Brave also supports blocking or allowing advertisements based on content types.

Devices can be compromised

In today's world, devices have many uses, with schoolkids to retired people all using devices for day-to-day activities. As well as individuals, most enterprises use devices to make their lives easy. Many users use multiple devices for different purposes. Devices range from workstations, laptops, and mobile devices to wearables, **Internet of Things** (**IoT**) devices, and assistant devices such as Alexa. These devices are primary targets for attackers. Even enterprise systems can be compromised through end devices, also known as **endpoints**. Due to this, a new approach was introduced for cyber security known as **endpoint security**. This protects endpoints from a range of attacks. The following list shows devices categorized as endpoints:

- Desktops
- Laptops
- Tablets
- Smartphones
- Servers
- IoT devices

- **Point of Sales (PoS)** devices
- Digital printers
- Smartwatches
- Assistant devices

Endpoints are important to monitor, as they can increase the attack surface and be an entry point for a corporate attack. Endpoints are easy targets for attackers, as users directly interact with endpoints. Information gathering and malware attacks are especially common at endpoints.

Let's look at some of the common endpoint attack types:

- Theft of credentials
- Phishing and social engineering
- Malware attacks
- Stolen devices
- Ransomware attacks
- Compromising devices

Since the endpoint devices are smaller in size compared to typical devices, they are prone to physical theft quite easily. Once stolen, attackers can obtain information stored on the device and exploit the information if the device is not encrypted.

As we discussed earlier, malware attacks are another common way of compromising endpoint devices. Apart from malware, attackers use password dumpers to copy and steal saved passwords on systems. Once initial access is gained, an attacker can pull credentials from an endpoint. If it's a Windows endpoint, the attacker can try to collect credentials from the following locations:

- **Security Accounts Manager (SAM)** file: This file is a self-contained database that exists in Windows systems from Windows XP onward. SAM databases authenticate users, locally and remotely, who provide credentials that match the credentials on the database. Many attackers steal this SAM database and try to decrypt it using multiple methods, including rainbow tables (tables with precomputed hashes). Attackers can dump this file using the `fgdump`, `samdump`, and `pwddump` tools.

- **Local Security Authority (LSA)**: LSA is a service that manages authentication on a Windows system and a local security policy. When this service is running and active on a system, attackers can dump LSA secrets from the memory and acquire sensitive information from the memory dump. They create a memory dump from the **LSA Subsystem Service (LSASS)** from the target system.

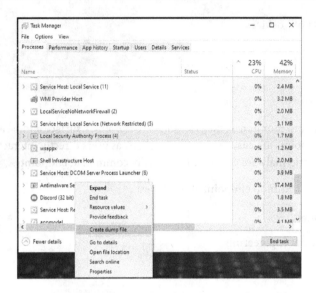

Figure 6.13 – Creating a memory from the LSA process

The preceding figure shows the creation of a memory dump by an LSA process on a Windows system. Once the memory dump is created, attackers have multiple ways to steal the credentials. The most common way is by using mimikatz.

The following figure shows how an attacker retrieves credentials from a memory dump created by the LSASS process:

```
  .#####.   mimikatz 2.2.0 (x64) #19041 May 19 2020 00:48:59
 .## ^ ##.  "A La Vie, A L'Amour"  (oe.eo)
 ## / \ ##  /*** Benjamin DELPY `gentilkiwi` ( benjamin@gentilkiwi.com )
 ## \ / ##       > http://blog.gentilkiwi.com/mimikatz
 '## v ##'       Vincent LE TOUX          ( vincent.letoux@gmail.com )
  '#####'        > http://pingcastle.com / http://mysmartlogon.com   ***/

mimikatz # sekurlsa::minidump lsass.dmp
Switch to MINIDUMP : 'lsass.dmp'

mimikatz # sekurlsa::logonPasswords full
Opening : 'lsass.dmp' file for minidump...

Authentication Id : 0 ; 101116 (00000000:00018afc)
Session           : Interactive from 1
User Name         : pentester
Domain            : DESKTOP-KFT86EI
Logon Server      : DESKTOP-KFT86EI
Logon Time        : 9/13/2020 5:50:03 PM
SID               : S-1-5-21-1984225036-1863897839-3139346863-1001
        msv :
         [00000003] Primary
         * Username : pentester
         * Domain   :
         * NTLM     : 92937945b518814341de3f726500d4ff
         * SHA1     : e99089abfd8d6af75c2c45dc4321ac7f28f7ed9d
        tspkg :
        wdigest :
         * Username : pentester
         * Domain   : DESKTOP-KFT86EI
         * Password : (null)
        kerberos :
         * Username : pentester
         * Domain   : DESKTOP-KFT86EI
         * Password : (null)
        ssp :
```

Figure 6.14 – The mimikatz dump password hash from the LSASS memory dump

- **WDigest**: Digest authentication is another authentication protocol used in Windows systems for **Lightweight Directory Access Protocol (LDAP)** and web authentication. This is a challenge-response protocol that keeps a plain text copy of the current user's password when enabled. Even though this protocol is disabled by default on the latest Windows systems, attackers can still enable this to steal credentials.

- **NTDS folder**: This folder can be located in the Windows domain controller, which holds a local database of domain objects. If an attacker gets access to this directory, it is not that hard to acquire users, groups, and credential information of the domain.

- **SYSVOL folder**: This folder contains group policy configuration-related information, including what types of policies are enforced in the domain and configurations of the policies. For example, if a password policy is enforced, an attacker can see the minimum length of the password enforced, the maximum age of the password, and the minimum age of the password, which the attacker can fine-tune with password-cracking tools accordingly.

Ways that an attacker can access your networks

Compromising networks is another common avenue for attackers to access systems. Typically, networks provide connectivity to other entities – for example, your home network that connects you to the internet. Typically, ISPs connect you to the internet over a modem. In today's world, the most commonly used network system for domestic requirements is Wi-Fi networks. There can be multiple devices connected to the same Wi-Fi network. Compared to wired networks, Wi-Fi networks have fewer risks.

Mainly, the coverage of a Wi-Fi network can go beyond your house or your premises. Wi-Fi networks use radio frequencies as a medium for transferring data. When you implement a Wi-Fi network, if the network range goes beyond your premises, anyone outside your premises can receive signals. Also, they can collect a lot of information, including the MAC address of the Wi-Fi device, the MAC addresses of the devices connected to the network, the number of devices connected, and active devices.

First, the attackers need an external Wi-Fi adapter to start Wi-Fi cracking; the reason for this is that most of the tiger boxes (operating systems with security and hacking tools installed for penetration testing), such as Kali Linux, Parrot Security, and Security Onion, run as virtual machines. For an attacker to crack Wi-Fi credentials, they need to monitor the signals first. In order to convert a Wi-Fi connection to monitoring mode, most of the cracking tools require a USB Wi-Fi adapter, although there are a few adaptors that are supported by these cracking tools.

The following is a list of Wi-Fi adapters fully compatible with Kali Linux, Security Onion, and Parrot Security:

- Alfa AWUS036NHA
- Alfa AWUS036NH

- Alfa AWUS036ACH
- Panda PAU06
- Panda PAU09
- TP-Link N150 TL-WN722N

Attackers will connect these Wi-Fi devices and start monitoring the networks. Some attackers even use high-gain antennas to receive signals from long distances. Once they start monitoring, they can collect information about targets.

Attackers use a range of techniques and tools to connect to target networks and gain access to them. Mainly, they use open source distributions such as Kali Linux and Parrot Security to crack target networks. They listen to the networks silently and collect information about the networks.

The following are some of the most used Wi-Fi cracking tools:

- Aircrack-ng
- Fern Wifi Cracker
- Kismet
- inSSIDer
- CoWPAtty
- AirJack

Once the external USB Wi-Fi adapter is connected, attackers can put the Wi-Fi adaptor into monitoring mode and monitor all surrounding Wi-Fi networks.

Figure 6.15 – Using the Aircrack-ng suite attacker to monitor all Wi-Fi networks within range

When an attacker starts monitoring, without even connecting to any of these networks, they can see a list of Wi-Fi networks within range, the signal strength, the number of data packets transmitted, the speed, the encryption type, the authentication type, and associated devices on each Wi-Fi network. From the list of surrounding Wi-Fi networks, the attacker selects the network they want to compromise and specifically collects information from the required network. When selecting a Wi-Fi network, the attacker must select from an active network, as they use data communicated between the access point and the connected devices to compromise the Wi-Fi network, depending on the encryption they use. Commonly used encryption types include **Wired Equivalent Privacy (WEP)**, the oldest encryption protocol, which is less secure compared with the other available Wi-Fi encryption and is rarely used now; **Wi-Fi Protected Access (WPA)**, which provides better encryption and user authentication than WEP; and **Wi-Fi Protected Access 2 (WPA2)**, which provides better encryption, using the industry-standard **Advanced Encryption Standard (AES)**. Recently, WPA-3 has been released, but most devices still connect using WPA2 encryption.

Depending on the encryption that is enabled on a network, attackers use different techniques to crack the Wi-Fi passwords of a target system. Once they crack the Wi-Fi password, they also can connect to the network and see the other connected devices and their communication. Sometimes, users think that when a network range is limited, attackers cannot access the network, as Wi-Fi signals are low. However, attackers can come equipped with high-gain antennas that allow them to receive signals from longer distances than users expect.

If a target network uses WEP encryption, attackers will try to collect a large number of communicated data packets (about 10,000 packets), and with the help of a wordlist, they can recover a Wi-Fi password. But when WPA2 encryption is used, an attacker can only find out the password using a four-way handshake. This is communication that happens between access points and a device before establishing a connection. The client device must produce an already configured key to the access point to authenticate. The attacker tricks this communication and intercepts the four-way handshake, which has information about the Wi-Fi key. Later, the attacker can crack this key offline using a wordlist or brute-forcing.

If attackers use `aircrack-ng` to crack Wi-Fi passwords, they open the terminal and execute the following commands:

- `Iwconfig`: When executed, this command will show attackers the number of network interfaces available including external Wi-Fi adaptors.

- `ifconfig wlan0`: When executed, this command will show information about the `wlan0` interface; if it is different, you need to change the interface accordingly from the information collected from the previous command.

- `airmon-ng start wlan0`: This command creates a monitoring mode using the `wlan0` Wi-Fi interface.

- `Mairodump-ng mon0`: When executed, this command will display the available Wi-Fi networks in the range. Then, attackers can select which network they want to compromise.

- `airodump-ng -w ourfile --bssid 58:6D:8F:XX:XX:XX mon0`: This command starts writing data to a file called `ourfile`, but only from the specified network access point (`bssid` specifies the MAC address of the access point, collected from the previous command).

Cloned session

While an attacker captures data communicated from an access point, they can start another terminal to initiate a four-way handshake so that it will be captured and written to the file by another running terminal. The following command is executed on a different terminal while the other terminal is running:

```
aireplay-ng --deauth 5 -a 58:6D:8F:A0:5B:16 mon0
```

This command sends de-authentication (a request to perform a four-way handshake) requests, spoofing the attacker's MAC address as an authentic access-point MAC address. Once this is successfully performed, other devices that are connected to the Wi-Fi network will perform a four-way handshake with the access point and be captured as `ourfile`.

When the attacker captures the four-way handshake successfully, the last step is to decrypt the four-way handshake to find the Wi-Fi key. This is performed by `aircrack-ng`, as shown here:

```
aircrack-ng ourfile-01.cap -w /pentest/passwords/wordlists/
darc0de.lst
```

This command will crack the Wi-Fi password from the captured `ourfile-01.cap` file with the help of a word list at `/pentest/passwords/wordlists/darc0de.lst`, or the attacker can use their own wordlist files. There are also free online services available to decrypt a four-way handshake once it is captured to a file.

If communication is over unencrypted protocols, attackers can sniff the network and collect information from the communication. Typically, HTTP, FTP, Telnet, SMTP, and SNMP are unencrypted protocols that communicate using unencrypted plain text. If any credentials are sent over a network, attackers can collect credentials easily. Since Wi-Fi networks with modern protocols don't broadcast information, attackers will not receive all the communication, even if they are connected to the target network. Then, attackers use **Address Resolution Protocol** (**ARP**) poisoning attacks to collect desired information from the target computer.

One of the most commonly used tools is Cain & Abel, which can perform ARP poisoning and collect credentials sent over a network.

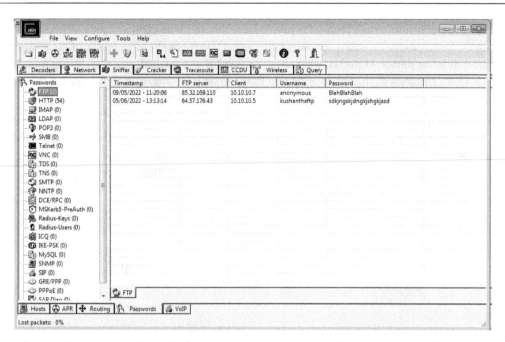

Figure 6.16 – Using the Cain & Abel attacker to capture data sent through unencrypted protocols

Now that we have seen the ways attackers can access our networks, let us see more details about how our information can be compromised by browsers.

Compromising browsers

The second topic of this chapter discussed how cookies are stored in the browser and can collect your information, especially the way third-party cookies will collect information about you. Apart from how cookies collect your information, your browsers keep heaps of other information. Importantly, if an attacker gets access to your computer, they can access and review your online habits, behaviors, and sensitive information. As well as attackers, many web applications can collect this information, using third-party cookies.

Let's look at the information that a browser reveals about you:

- **Web history and cached data**: Most of the popular browsers provide a **history** feature. This feature is important if you accidentally or deliberately close a web page and forget the URL – it can be your savior. On the other hand, think about privacy. As the browser history keeps information about every single web page you visited, if anyone gets access to your device, it can reveal every single web page you have accessed.

Most browsers today create profiles for you. At face value, this is an exciting feature, as irrespective of the device, users can access the same information if they use the same profile of the browser on a different device. However, what about privacy? If any single device is compromised, an attacker can access all the information that the browser is storing, even if the user has deleted it from the local browser.

Let's take Google Chrome as an example. Since Chrome is from Google, users don't have to create an additional profile or account; they can use their existing Google account instead. If the user is logged in to their Google account while using the Chrome browser, it keeps track of all the information in their Gmail profile. You can try this by accessing `https://myactivity.google.com/`:

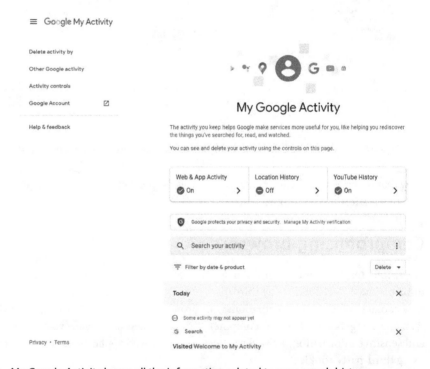

Figure 6.17 – My Google Activity keeps all the information related to your search history

This shows how much data Google collects from users and keeps within their profiles. If you analyze a single search item that is saved in the **My Google Activity** section in your Google account, you will understand more about how much data that browser can collect. From the items that are listed under **My Google Activity**, let's click on the **Details** link on an item and see how much information Google keeps about that item.

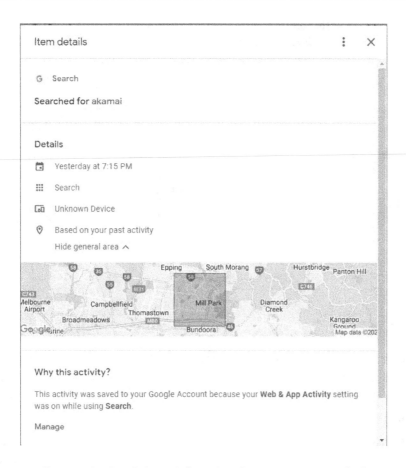

Figure 6.18 – Google keeps information about every item searched

As the preceding screenshot shows, Google keeps track of the search keyword, the type of search, device information, and the location of each search item.

As well as Google Chrome, the most popular browsers do the same. They also keep track of a lot of information about each search that you performed.

Browsers also keep the cached information. Cache refers to the contents that are stored in the browser from previously accessed websites. A cache provides faster access and loads web pages much faster. But from a privacy point of view, if anyone has access to the device, they can also access a cache quite easily and acquire information that you would not want anyone to see. To access a cache, you can type chrome://cache in the Google Chrome browser address bar or about:cache in Mozilla Firefox.

There are some tools that retrieve cached information from browsers. You can download some tools from `https://www.nirsoft.net/` to retrieve cached data. The following are some of the tools that can retrieve cached data from browsers:

- **ChromeCacheView**: Retrieve cached contents from Google Chrome:

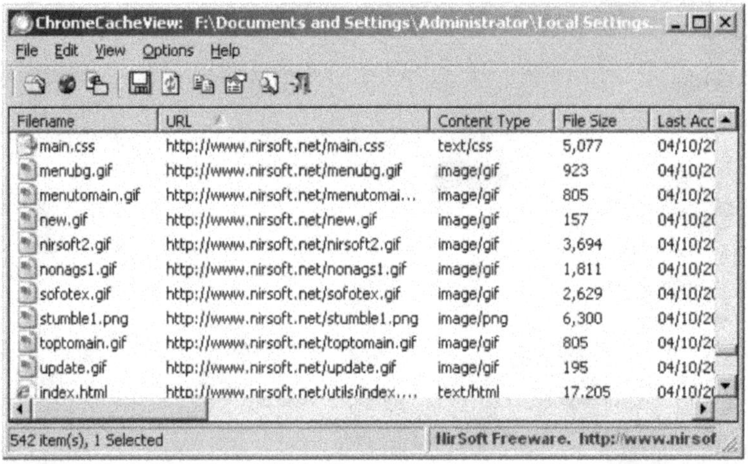

Figure 6.19 – ChromeCacheView to retrieve cached information from Google Chrome

- **IECacheView**: Retrieve cached contents from Internet Explorer.

- **MZCacheView**: Retrieve cached contents from Mozilla Firefox.

- **VideoCacheView**: If you watch a video on a browser, this utility can automatically search videos cached in Chrome, Firefox, Internet Explorer, and Opera and can save the video to another folder.

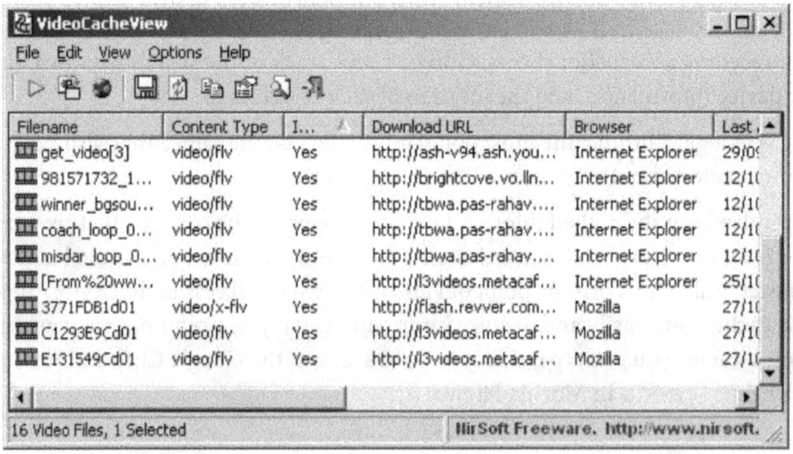

Figure 6.20 – VideoCacheView can search videos cached in various browsers

From `https://www.nirsoft.net/`, you can download and test a range of tools that can be useful to retrieve information from browsers. Even though these tools are GUI-based tools, they also support command-line scripting. Attackers use these tools to alter the scripts or create AutoRun files (when attackers create AutoRun files, they will automatically get executed when an event is triggered) that will automatically execute and write important information to a text file.

For example, **MyLastSearch** is a small free utility that can be downloaded from the same site that scans the cache and history files on the target web browser, and it can also collect your search queries executed in popular search engines, such as Google and Yahoo. If you want to include this in a script, you can easily write a script with the following:

- `MyLastSearch.exe /stext <FileName>`: This will save the located search queries in a text file.

- `MyLastSearch.exe /scomma <FileName>`: This will save the located search queries in a comma-separated file.

- `MyLastSearch.exe /shtml <FileName>`: This will save the located search queries in an `html` file.

- **Cookies**: Cookies collect information about the user, devices, and other information, as we discussed in the second topic of this chapter.

- **Bookmarks**: You may be wondering what an attacker can do with your bookmarks. If you access the bookmark manager built into your browser, you will realize that it not only keeps shortcuts to your financial websites, banks, and insurance websites but also very specific login information. For example, when you save a website login as a bookmark, some web applications generate a very specific URL for you to log in to the application securely. When we save it as a bookmark, we save the special link generated by the web application as a shortcut. If an attacker discovers you are with a specific bank (which can be collected from your bookmarks or favorites) and your email address, it's just a matter of stealing a password, which can be easily done by creating a phishing website, as the attacker already knows the bank that you are with. Let's go to the bookmark manager of the browser and click on any of the saved web logins, go to **More Actions** (or right-click), and then click **Edit**:

Figure 6.21 – Bookmarks keep special URLs generated by web applications

As you can see from the preceding screenshot, this user has saved a financial institution's login URL, but when the user saved the login, it also saved a specially crafted URL generated by the web application. This could be to validate the device or a trustworthy user. If an attacker uses the same link, the application might accept the attacker as a trustworthy device or user.

- **Browser extensions**: Most of the popular browsers support extensions to enhance their capabilities. These extensions were created by companies and individuals. Most of the extensions are legitimate, but can we be sure that they all are? As extensions are connected to a browser, they also have access to most of the functions of the browser. Even without our knowledge, these extensions can collect sensitive information.

Most of the extensions are free, but we never know what the real intention of the extension is. It might look like a very useful extension, but under the hood, what does it really do?

Microsoft SysInternals has an interesting set of freeware for different purposes. If you download TCPView, which is a lightweight free tool, it will show you what the connections are that were created by your device. If the process creates any connections from your browser to suspicious remote addresses, you can monitor using TCPView.

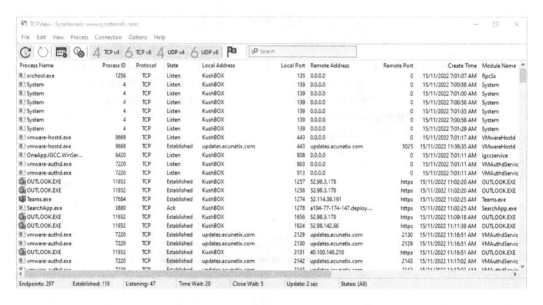

Figure 6.22 – TCPView will show the connection information

This is a useful tool to monitor the connection status from a device to remote addresses. This tool shows the list of current processes and the connections created and initiated by different processes. If the browser extension creates suspicious links to remote addresses, you can review them using this tool.

The best practice is not to use any of these extensions. But if you really want to use an extension, it is better to assess and monitor before you start to use any extension, as we never know the real intention of the developer. As discussed, most of the extensions are legitimate, but there can be extensions developed with malicious intent.

Summary

This chapter focused on the different artifacts that attackers collect from target users and companies. As well as attackers, competitors collect large amounts of information from you. Attackers target devices to collect information using different approaches. Devices collect and keep various information about users and their activities. When a device is compromised, attackers will have access to this information. We also looked at the ways attackers compromise networks. Finally, we discussed the ways attackers compromise browsers and the types of information that can be collected from the browsers.

In the next chapter, we will explain what cyber anonymity is and its different levels, as well as the difference between privacy and anonymity. We will also explain the best practices to maintain cyber anonymity. Cyber anonymity is not something that you might have to do only once, but you need to develop a cyber-anonymity mindset to properly maintain cyber anonymity and face the various types of privacy-related attacks.

Part 3: Concepts and Maintaining Cyber Anonymity

This part explains the concepts of cyber anonymity and how to maintain it.

This part comprises the following chapters:

7

Introduction to Cyber Anonymity

In the previous chapter we talked about the artifacts that can be collected by an attacker from individuals and companies from devices, networks, and browsers. Not only attackers but also competitors collect information and artifacts from users. We discussed the tools and techniques that attackers use to compromise networks and we discussed how browsers can be used by the attackers to collect information from users in particular.

This chapter we will be focusing on cyber anonymity. The flow of this chapter will be as follows:

- Definition of cyber anonymity

- Privacy and anonymity

- Levels of cyber anonymity

- Best practices to maintain cyber anonymity

- Developing a cyber anonymity mindset

Definition of cyber anonymity

The term **anonymity**, like many other English terms, originates from two Greek words. The first half of the word, **an**, is derived from a Greek word meaning **without**, and the second half, **onoma**, is another Greek word meaning **name**. Combining these two words created the term **anonymous** – maintaining the state of going unnamed is referred to as **anonymity**. When it comes to the cyber world, anonymity refers to remaining online without revealing your identity. There are different levels of anonymity, which will be discussed in detail in the third topic of this chapter.

The moment you connect to the internet, your information will be collected in many layers, as we discussed in earlier chapters. The meaning of cyber anonymity is making it impossible for others to find the owner of a message or an act by not revealing their identity while remaining connected.

Another definition of cyber anonymity is performing activities connected to the internet without your identity being revealed. As you can see, anonymity is also associated with privacy. When privacy is missing, anonymity cannot be maintained. Whoever would like to maintain cyber anonymity has to protect their privacy first. We use different identity systems in physical interactions, including national identity cards, passports, driver's licenses, and social security numbers, but in the cyber world, mainly, identities are represented by usernames or user IDs.

On the other hand, being anonymous considerably reduces accountability for any actions performed. This is the main reason for attackers to take precautionary actions to hide their identity when performing illegal acts, which makes it hard or impossible for investigators to trace the perpetrators after the action.

Because of this, it has become a legal requirement for many online services to provide an identity before performing any action. Once an identity is provided, it must be verified, as attackers will impersonate the identity otherwise. There are many different identity verification methods. All the identity verification methods are categorized into six categories. Identity verification is also known as **authentication**. The following are the commonly available authentication methods:

- **Knowledge-based authentication** (**KBA**) – KBA verifies user identities by requiring knowledge-based information. This can simply be a password, a PIN number, or a series of questions that only the user is expected to know. These questions are not very common questions – for the user, it's easy to answer but very hard for others. This authentication usually hardens with a time limit for the answer and attempt limit. Most importantly, KBA is the easiest method for users to use to authenticate themselves, but the disadvantage is that this can be compromised by attackers gathering information about the target. It's the user's responsibility not to reveal sensitive information to the public.

- **Multi-Factor Authentication** (**MFA**) – MFA or **Two Factor Authentication** (**2FA**) is an authentication method where the user must go through two or more authentication steps by entering a password and a code received on their mobile phone. This authentication uses any two or more of the following authentication types:

  - Something you know – something you know to verify your identity (for example, a password or PIN)

  - Something you have – something that the user possesses that can verify identity (for example, a smart card, a mobile phone to receive the code, or a digital key to verify your identity)

  - Something you are – biometric authentication to verify your identity (facial recognition, fingerprints, iris scanning, or retinal scanning)

  Since this verification is commonly used, users generally have an understanding of how to provide MFA or 2FA authentication. This requires users to provide additional verification other than a username and password based on the verification options available.

- **Credit bureau-based authentication** – With this verification method, when the user requests access, the user will be verified based on their credit status. If you have a good credit history, access is granted – this information is collected from large credit databases. This verification method is used for financial-related authentication, but the downside is that user credit status is used as a verification method.

- **Database-based authentication** – Database-based authentication is another authentication system that validates the user with the information entered compared to the information stored in the database. For example, when a passport is submitted for renewal, the status of the renewal can be tracked by providing the passport number and the application date. This verification method is generally used to provide the status of an activity and is very easy to use.

The downside of this verification method is that another person can impersonate this by guessing and providing partial information. The following is an example of this verification method:

Infringement Search

Infringement Number

Vehicle Registration

Search | Clear

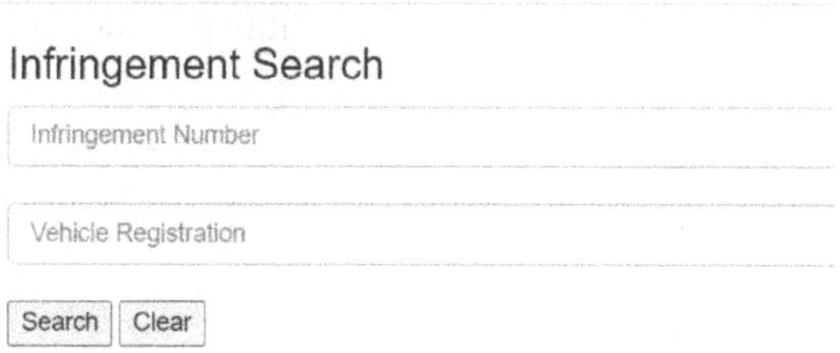

Figure 7.1 – Database-based authentication

Database-based authentication is used to authenticate by providing information that is already stored in the database. As per this figure, the user must provide both an infringement number and vehicle registration to authenticate and access the system.

- **Online authentication** – This method is employed by many web portals today, by integrating an external authentication provider. For example, many web applications and websites allow you to access them by providing Facebook or Google authentication. The downside of this method is the service provider does not have visibility into the authentication and someone could impersonate this authentication with fake social media accounts or stolen accounts. When the user needs to authenticate, the service provider redirects the request to the authentication provider. The user then enters the credentials at the identity provider's portal and authentication happens at the identity provider's portal:

Figure 7.2 – The service provider provides external authentication providers to authenticate

As per the preceding figure, `www.scribed.com` provides users with access to a range of books. Users have the option of creating an identity on `www.scribed.com` or they can use an existing Google or Facebook account. When a user selects an external identity provider, the request will be redirected to the respective identity provider's portal. Then, the user can provide the credentials:

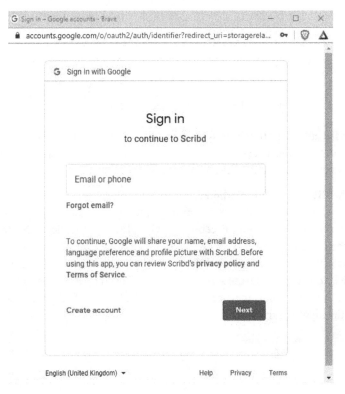

Figure 7.3 – When a user selects Google as a service provider, the request is redirected to Google

When the user provides valid credentials in the identity provider's portal and is successfully authenticated, the identity provider redirects to the service provider again to provide access to the service.

- **Biometric authentication** – As we discussed under MFA and 2FA authentication, biometric authentication uses biological attributes for authentication, which are not easy to impersonate. Biometric authentication can be part of MFA or 2FA or used as a standalone authentication method.

Typically, once the user is authenticated, the user will be given access based on the least permission policy, which is known as **authorization**. Most systems, including financial organizations, practice this mechanism. Once the user is authorized (authorization is the level of access given to the user based on the user role), the user will be given access to the resources based on their permissions. The permission will be given based on their role. According to the security policies, the minimum set of permissions that allows them to perform their job tasks will be assigned.

In today's world with its complex requirements, the preceding authentication systems cannot provide complete security, as attackers can use various tactics, techniques, and tools to compromise security. To overcome this concern, many identity systems incorporate zero-trust-based implementations, which are not only limited to the preceding authentication methods but also validate additional attributes such as the usual location of access, the usual device of access, realistic travel times, IP addresses, and suspicious behaviors.

For example, Azure Active Directory is one of the most used identity systems and provides identity services internally and externally. While supporting MFA or 2FA authentication and biometric authentication, it also supports configuring conditional access policies based on a range of criteria including the following:

- **User risk level** – If the user's credentials are compromised, commonly used or weak users will be categorized as high-risk users and we can configure a conditional access policy to restrict access for high-risk users even if the given credentials are correct.

- **Sign-in risk level** – If the user is logging in from a suspicious IP range or a user's login history is unrealistic (for example, the same user logged in previously from Singapore 15 minutes ago and now the user is trying to log in from the US), the user will be blocked from logging in even if authentication is successful.

- **Device platforms** – Restricts access based on the device platforms – users can be blocked from access when they are using Linux systems, for example, even if the provided credentials are correct.

- **Locations** – Allows or denies access from certain locations based on countries or IP addresses (excludes MFA when the user is connecting from a trustworthy location such as company headquarters).

- **Client apps** – Blocks or allows based on the app. For example, if the same app is available as a desktop app and a browser-based app, administrators can restrict users from logging into the system using browser-based apps.

- **Filter for devices** – Provides access based on device properties including device ID, display name, device ownership, manufacturer, model, operating system, operating system version, and many more attributes of the system:

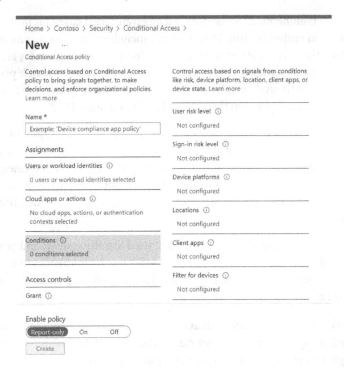

Figure 7.4 – Azure Active Directory provides a conditional access policy to configure zero-trust security

If the user identity is critical, systems and infrastructures can use conditional access policies in addition to main authentication systems. Conditional access supports maintaining zero-trust security. The idea of the zero-trust security implementation is not to trust any request without verification.

For example, when a user provides the correct credentials to log into the system and the system is designed to validate a user only based on the username and password combination, we are assuming that the user credentials will never be compromised. As a result of this implementation, if an attacker compromises the user credentials and tries to access the system, the system will allow the attacker access if the credentials are correct. In today's world, there are many attacks targeting user credentials. Compromising passwords has become common. If the systems are completely reliant on the username and the password, there can be many attackers accessing the system pretending to be users using compromised passwords.

Therefore, zero-trust implementation is necessary. Even if attackers successfully compromise a password, the system will still have to validate the user based on the different verification methods that we discussed before.

For example, if an attacker compromised a user's password, typically, the attacker would launch this attack remotely and then try to access the service; if the system is based on a typical castle security implementation, the attacker would be able to successfully access the resources. If the systems are configured with zero-trust security, a username and password would be not enough to access the system even if these credentials have been compromised. When an attacker tries to access the system using the compromised credentials, even if the credentials are correct, the system will try to validate the following:

- Whether the request is coming from the usual location
- Whether the request is coming from the usual device
- Whether the previous attempt to access the service and the current request attempt IPs are coming from the same region, and if not, whether the travel time is realistic
- Whether the current request is coming from a suspicious IP range – systems maintain real-time monitoring systems to collect suspicious IP ranges
- Whether the request is coming from known **Virtual Private Network** (**VPN**) systems commonly used by attackers

Since the system is validating the request based on the preceding criteria, even if the attacker is trying to access with compromised credentials, the system will block the attempt in real time, which makes the attacker's successful attack useless.

This section explained the importance of implementing a proper identity system, as cyber anonymity is based on not revealing your identity. Especially for critical systems, it is very important to maintain a proper identity system – not only systems authenticating based on credentials but also monitoring other attributes that make attackers' attempts much harder to succeed.

Privacy and anonymity

There are many arguments about privacy and anonymity – some of these arguments try to establish a connection between privacy and anonymity – but privacy and anonymity are two different concepts. In other words, maintaining privacy will not lead to anonymity. Depending on the scenario or the requirement, you might want to choose privacy or anonymity. Having a better and clear understanding of privacy and anonymity will help you to select the right option. For example, when using a mobile app, when accessing a web application, or when installing software, it will tell you that the app or software will maintain your privacy – or that it provides anonymity. Some organizations or companies are being honest here, but some are still playing with words, as most users do not have a very clear understanding of these terms.

What is privacy?

The term **privacy** refers to the ability to keep your personal or sensitive information exclusively to yourself and have total control over access to your information. In other words, you can control who can access your data, what the level of access is, and when they can access it and you can find out what the purpose of them accessing your information is. As a broader definition, information privacy is the right to have control over your information and how it can be collected, accessed, and used. This will often be dictated by the privacy policy in many organizations or when you are accessing any online service in the cyber world. The privacy policy is treated as a legal document that defines the way customer data is gathered, used, managed, and disclosed.

In the previous chapter, we discussed cookies – especially, we discussed how third-party cookies will collect information from users and often share it with other companies or organizations. As you may have noticed, many websites you visit today have cookie policies. Even though users frequently won't read it, the cookie policy defines the information that they gather and how the information will be used. As an example, let's visit the `https://www.packtpub.com` website. If this is your first time visiting this website using this browser, you will be prompted with a cookie policy acceptance notification.

This website has given users options to decide on collecting information. As it clearly says, "**This website uses cookies and other tracking technology to analyse traffic, personalise ads and learn how we can improve the experience for our visitors and customers. We may also share information with trusted third-party providers.**" Once the users give consent, their information will be collected. On the other hand, cookies will be used to provide a more personalized and rich experience for the users based on their choice:

Figure 7.5 – Cookie policy to get user consent on information collection

If you click **More info**, this website will take you to the privacy center where you can select what type of information collection you consent to. There are usually a few options: this website explains the privacy information, the cookies necessary for the website to function properly, which usually users cannot turn off, performance cookies, which typically do not collect information, and the third-party cookies that we discussed in the previous chapter:

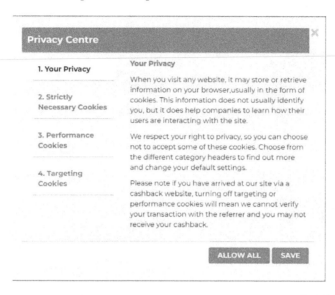

Figure 7.6 – Privacy statement on cookies

If you select the second option, **Strictly Necessary Cookies**, it will not allow the user to perform any action or turn off cookies that will affect the site functionality.

When you select **Performance Cookies**, you can allow or disallow cookies to be active. Performance cookies collect performance-related data but not personal information about the users. For example, these cookies collect information such as how many users visited this website, how long they stayed on the website, and the number of pages visited. You can select whether you want to enable these cookies or disable them. The fourth option, which is **Targeting Cookies**, is cookies from third-party providers. In particular, these cookies will collect information and may share it with third-party providers. This will be used by third-party providers to personalize advertisements.

If you click on the **Learn more** link of this website, it will take you to the company privacy policy, which explains how they collect customer data, including the information collected through cookies, what type of data they are collecting from customers, how they manage the data they collect, how long they will keep the data, with whom they will be sharing this information, and where the personal data will be processed:

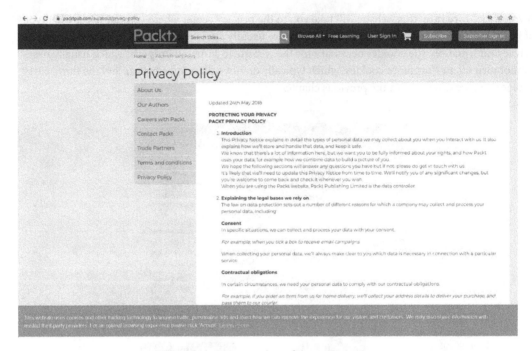

Figure 7.7 – Privacy policy of the organization

If you read through the policy, you will find out how they collect the data. For example, they will collect data when you visit their website, when you make online purchases, when you engage with their social media, when you install their app, or when you create an account with them. They explain what type of information they collect from the customer, including customer name, date of birth, billing information, job title, telephone number, and billing information. They also explain why they collect customer data and how they manage customer data.

This example shows how this website collects customer data, manages customer data, and uses customer data. The customer has the option to decide what type of data they can collect.

What is anonymity?

Anonymity refers to hiding your identity but not your actions. As we discussed before, when you interact with the cyber world, different entities will be collecting your information in various layers. In the cyber world, you can be anonymous by preventing these entities from collecting your identity-related information. It's more like in the physical world when someone is covering their face to prevent others from identifying them. In many robberies, robbers used to wear face masks to prevent others from identifying them. In that way, they could be anonymous, but their actions were still visible. For example, if the robbers rob a bank and wear face masks, the action is still visible, as many have seen the bank robbery take place, but they cannot identify the robbers, as they were wearing face masks. In many cases, investigators are able to catch the robbers by using a tiny bit of information that can still identify the robbers. Sometimes, it can be a tattoo on one robber's hand, for example, which can be used by the investigators to trace the person even though they were wearing face masks.

In the same way, in the digital world, there are ways that you can be anonymous, but a small mistake can reveal your real identity. Attackers wanted to maintain anonymity all the time, as their intentions are bad and they never want to get caught. As users, our intention to be anonymous is based on ethical reasons, as we don't want our information to be collected and processed without our consent.

Since user data will be collected in multiple layers and multiple entities, it's not simple to be anonymous in the digital world, especially for users. When it comes to attackers and hackers being anonymous, it's not that hard, as they are aware of these layers – as in, the ways that different entities collect identifiable information. However, for users, remaining anonymous is challenging. This is the reason we discussed various types of data collection methods. When you understand the ways that the information is collected, then you can understand and plan to prevent your data from being collected at various layers.

In the next few sections, we will be talking about different levels of cyber anonymity and the best practices when it comes to cyber anonymity. If you really want to maintain anonymity when you are interacting with the cyber world, developing the required mindset is very important. As I explained earlier, through a single mistake, investigators can trace a robber; the same thing can happen when you neglect a single piece of the process and reveal your identity.

Levels of cyber anonymity

As we discussed earlier, cyber anonymity is trying to hide your identity without hiding the action. The meaning of being anonymous is hiding your identity but your actions still being visible. Back in 1996, there was a paper published in the Journal of Universal Computer Science by Bill Finn and Hermann Maurer, who were from the computer science department at the University of Auckland. It first introduced the levels of anonymity (https://www.jucs.org/jucs_1_1/levels_of_anonymity/Flinn_B.pdf). This paper introduced multiple levels of anonymity. According to the paper, networked computer systems required multiple levels of anonymity. The paper explains five levels of anonymity, but these levels were introduced in 1996, so it does not provide categorization for the techniques and tactics used today. However, it establishes a few points to continue our discussion.

The various levels of cyber anonymity are as follows:

- **Level 5 – Super-identification** – According to the explanation in the paper, this level completely identifies the user and authenticates the user using the user ID and the password to the system in a unique way. All the communication transactions carried out by each user will be stored as an audit trail for later investigation if required. In those days, enterprise systems with mainframe computers used this secure identification system. When you compare this idea with today's technology, it's pretty much like the zero-trust authentication we discussed. In those days, it applied to mainframe systems with closed environments (closed infrastructures are not exposed to other networks or the internet).

- **Level 4 – Usual identification** – This is explained as systems that totally rely on a username and password combination only. If anyone has the correct username and password, the system will allow the user to access the system and access resources without validating other attributes. If you compare *usual identification* to today's systems, this is like systems that authenticate users only based on credentials. We discussed castle security and zero-trust security before. If an attacker compromises the username and the password, they can access the target system without any problem, as the system only validates the user based on the username and password combination.

- **Level 3 – Latent or potential identification** – In this identification system, users use pseudonyms in the system. Each user has a pseudonym and is mutually disjoined, which means each user will have a screen name or username in the system, but one user cannot identify another user in the system in a real, personally identifiable way. As a result, two users cannot identify the other user's identity directly. While the system has complete knowledge of each user, user-to-user communication is always pseudonymized. When you compare this type of identity with today's scenario, it's mostly in community discussions, technological forums, and bulletin board discussions that people use stage names and pseudonyms to introduce themselves. You can only identify users by the stage name or commonly used profile name. You can also find this on social media such as Twitter, YouTube, and TikTok – many people use profile names, not their real identity. Some profile users can see the person so they can identify the person, but on many profiles, they maintain some level of anonymity outwardly while providing their real identity to the system.

- **Level 2 – Pen-name identification** – This identity system is a bit like latent or potential identification, as users can use pseudonyms, but the difference is that even for the system, their real identity is not visible. With latent or potential identification, even though the user's real identity is not visible, the system has complete knowledge about the user. In pen-name identification systems, even the system does not have complete knowledge about the users. The user can create an account by providing an email address – if the email system can be used to communicate and activate the account, the user can create an identity using a pen-name identification system. Since many systems provide free email accounts, users can create an email address without properly identifying themselves; thus, these systems do not have complete knowledge about the users. To apply this level of identification to today's world, some online gaming platforms use pen-name identification, as do some community and discussion boards.

- **Level 1 – Anonymous identification** – In this identification system, users are identified by the system but not as addressable users. No username is required. Neither the system nor the other users can uniquely identify the user. The system keeps logs of events related to this entity and its activities but is not able to distinguish the user, mainly to tailor system interactions based on user activities. On examining this identification, I would recognize this as *partial anonymous identification*, not fully anonymous identification. When comparing this to today's technology, this is very similar to the technique we discussed in the previous chapter related to direct cookies. Direct cookies keep information about the user without an identity system.

Direct cookies collect information related to the user's device, browser, location, or IP address, but cookies will not be able to identify the user by a username.

- **Level 0 – No identification of user** – This identity system does not require any identity or password combination at all. Users can access the system without user IDs. The system still collects information about the user activities, as that's how the systems are designed, but cannot distinguish the user by username or any other identity mechanism whatsoever. I would like to refer to this as *partial anonymous identification*; the reason is that even these systems collect information about the user's behavior and keep information that can be used to tailor interaction with the user. This is like what third-party cookies do today. They don't uniquely identify the user, but they collect data related to the user behaviors and search for patterns that will help them to send tailored advertisements – they can share these with other providers and these can send similar advertisements. According to the paper, this provided the highest level of anonymity at that time. Now, there are newer and better technologies introduced to maintain a better and higher level of anonymity.

The reason to start our discussion based on this paper is that this paper establishes the ground to continue our discussion of further levels of anonymity. As you may notice, Level 1 and Level 0 discussed in Bill Finn and Hermann Maurer's paper do not provide proper anonymity, as they collect information about the user, even though they don't collect identity-related information. Since this paper was published two decades ago, we would need to achieve a greater level of anonymity in today's complex systems.

Beyond Level 1 and Level 0

According to the paper we discussed, the highest levels of anonymity were provided by Level 1 and Level 0, but as we understood, even though both Level 1 and Level 0 do not collect the identity or do not require authentication, both implementations collect user activity-related information. The reason behind this categorization is the definition of anonymity. By definition, anonymity hides identity, not actions. Since Level 1 and Level 0 do not collect information related to identity, the paper presented by the University of Auckland defined even Level 1 as an anonymous system.

When we compare this situation with today's world with more complex implementations, collected information can contain sensitive and personally identifiable information even though the user identity is not collected. For example, as we discussed, direct cookies and third-party cookies collect information related to the device, browser, location, user behavior-related information, IP address, and any items that the user is interested in. Combining all this information, you could probably uniquely identify the user. If we want to establish another level beyond Level 1 and Level 0, we need to look at a system or method where none of this information is collected from the user, including the user's device, browser, IP, or anything related to the user's activities. If we suggested a system that did not even collect this information, it could be named a super-anonymous level.

Super-anonymous level

If we were to implement a super-anonymous level, mainly, it shouldn't have any identity or authentication requirements as per the definition of anonymity. Then, the real challenge would be to protect users from the systems established to collect user activity-related information, as we discussed under Level 1 and Level 0. Since most of the applications developed today use web-based technologies and are accessible over browsers, inherently, browsers use direct and third-party cookies to collect information. A super-anonymous level will be a level that does not collect identifying information or any other user device-based, browser-based, or behavior-based information during the interaction with the web-based application or website. Ideally, when accessing the system, it should not only avoid collecting identification-related information but also any activity-related or behavioral information.

To maintain a super-anonymous level, we need to follow the best practices to be anonymous on the internet. As discussed, user data is collected in multiple layers, so we need to follow best practices to prevent data collection when on the internet.

Best practices to maintain cyber anonymity

We discussed the layers of cyber anonymity and how the different entities collect user information while the user is on the internet. As this process is collecting information in different layers, we need to concentrate on all the layers, not just the browser. The best practices that we are going to discuss here not only concentrate on the browser but also all the layers. Let's look at some best practices to maintain cyber anonymity:

- **Using a VPN** – Whenever we connect to the internet, as we discussed, our IP can be collected by the respective web application or service. When you connect to the internet, there will be two IP addresses, called a private IP and public IP address. What they collect is our public IP address, which is assigned to you and is unique. This means there cannot be two devices that have the same public IP address at the same time, which means it will be unique. If you type `ipconfig` into your terminal window, it will show your private IP:

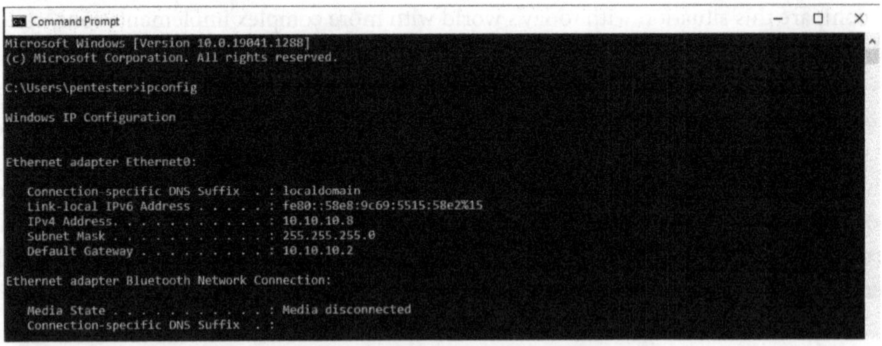

Figure 7.8 – This shows your private IP address

This will show your private IP address. In this system, the private IP address is `10.10.10.8`. If you want to check your public IP address, there are multiple ways to do that. The easiest way to check your public IP is by accessing `https://ip.me` or you can just search `whats my ip` on Google:

Figure 7.9 – The ip.me site shows your public IP address and other information

When you access `https://ip.me`, it will show you your public IP address and other information including your internet service provider, country, location, and postal code. If you are using a VPN service, you can send your traffic over a VPN server and this will prevent the web application from detecting your public IP. There are different types of VPN services available, which we will be discussing in the next section. For now, I will use OpenVPN to show you how the traffic is sent over the VPN server. Let's download the OpenVPN community edition client first by accessing `https://openvpn.net/community-downloads/`, and once downloaded, install the software onto our device. Then, we need to download the configuration file – we can download many configuration files on `https://www.vpnbook.com/`. There are many connectivity details available, but select OpenVPN, as we need connection details for OpenVPN:

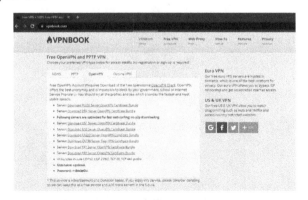

Figure 7.10 – Configuration files for OpenVPN for different servers

Select any of the listed servers to download the configuration file and extract it to any folder. Then, open the OpenVPN community edition software and import the files. Just select the **FILE** option and click **BROWSE** to select the configuration file's location or you can simply drag and drop the files:

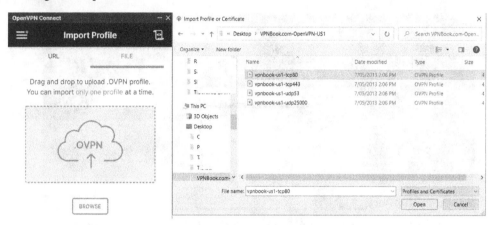

Figure 7.11 – Importing configuration files for OpenVPN for the servers

Once the configuration file is imported, you can specify the username provided by the https://www.vpnbook.com/ site and click **CONNECT**:

Figure 7.12 – Connecting to the VPN server

Once you click on **CONNECT**, it will prompt the password. You can provide the password provided by the `https://www.vpnbook.com/` site when you downloaded the configuration file:

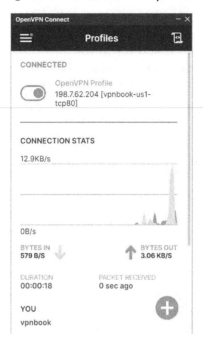

Figure 7.13 – Connected to the VPN server

Once the VPN client is successfully connected to the server, your traffic will be redirected through the VPN server. If you access any internet service now, they will be detecting the VPN server's public IP address instead of your public IP. Now, when you try the same site, see the IP address shown there:

Figure 7.14 – Once connected to the VPN server, your public IP will be changed

Once you connected through the VPN service as shown, it will not disclose your real public IP; instead, it will show the VPN service-connected IP address. Traffic will be encrypted from your device to the VPN server. This ensures anonymity while you are interacting with the internet.

Another important thing to remember is that the VPN is only connecting you through its server, so the VPN service should be trustworthy. If you use an untrusted VPN service, that VPN service provider can also collect your information. Some browsers and operating systems have VPN services built in. The Tor browser is the best example, which we will discuss in detail later.

- **Use a secure device** – Since devices collect information while we are interacting, if you want to maintain cyber anonymity, you need to use a secure device. The meaning of a secure device is a device that you have total control over. If the device is shared, there are multiple people using the same device, so they can collect your data, but when you are using your own device, you can keep information to yourself, or you can even use virtual appliances to make it more secure, which we will discuss later. Once you use your device, you have the freedom to remove any data or logs created on your device. To make it more secure, you can encrypt the device using a strong encryption algorithm. You can use either BitLocker or VeraCrypt to encrypt the device.

  VeraCrypt is a free and open source tool that uses strong AES256 encryption to encrypt data – you can use VeraCrypt to completely encrypt a device, including its operating system.

  BitLocker also supports the **Trusted Platform Module (TPM)**, a hardware module that provides trusted technology to protect sensitive data, which provides encryption much stronger and that's connected to the hardware device. When stronger encryption is used and the TPM is enabled, even if the device is stolen, a third party will not be able to retrieve the data.

- **Use a safe network** – When you are connecting to the internet, even the network connection collects information about you. Don't use free or public networks to connect to the internet, as you never know what they collect about you. When you are using your own trustworthy network connectivity with a stronger Wi-Fi password, at least you know that no one is sniffing your communication. If the Wi-Fi network password is guessable or weak, attackers can connect to your network easily and collect information using various techniques including sniffing. Using stronger encryption is important in the network – currently, WPA2 or WPA3 can be used as encryption to the WiFi network.

- **Using a secure browser** – We discussed secure browsers in the previous chapter. Brave can be a better option when compared with other browsers such as Chrome. Google heavily collects data from users and using Chrome exacerbates this. Using Tor is the best option, but for general usage, Brave will be sufficient.

- **Disable cookies** – We discussed this in the previous chapter in terms of important information, especially about direct cookies and third-party cookies – how the direct cookies and third-party cookies collect information and the types of information collected by them. We also discussed how we can disable third-party cookies on commonly used browsers along with the steps to do so.

- **Use a stronger password** – Even though we discussed many authentication systems, the username and password combination is still treated as the most used and convenient authentication method. When you are accessing your device, connecting to the network, and encrypting a device, you require passwords. When configuring passwords, we need to use super-strong passwords, as there are many password-related attacks. Many passwords are guessable and many passwords contain known information about you. There are methods to recover these passwords quite easily. Typically, a stronger password is a password that has more than eight characters with a combination of an uppercase character, a lowercase character, a special character, and a digit, but many users use passwords such as `Qwerty@123`, `Test@123`, and `Admin@777`, which fulfill the criteria required to become a strong password but can be commonly found in many password word lists.

If you use known information about you within the password, attackers use a method called **Common User Passwords Profiler** (**CUPP**) to generate a tailored wordlist generated to break your password. There are many tools you can find to generate CUPP passwords. I'm using `cupp.py` to generate this. You can download `cupp.py` from `www.github.com/Mebus/cupp`:

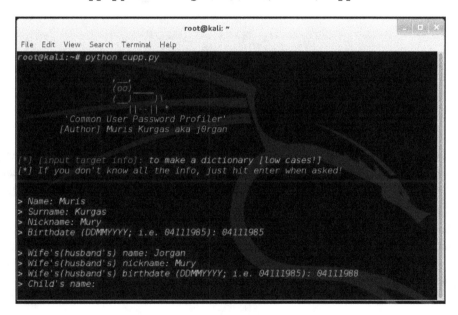

Figure 7.15 – CUPP generating a tailored password wordlist

`Cupp.py` is a Python script into which an attacker can enter known information about the user, including first name and last name, a spouse's name, a child's name, and birthdays. Then, the script will jumble the entered words, changing capital and simple letters and common special characters to create a long wordlist to crack the password.

Because of this, we shouldn't be using information about ourselves or family members within our passwords. One of the ways to overcome this issue is using passphrases, as phrases are easier to remember than complex passwords and they are lengthy. Therefore, they are not as easy to break using common password-breaking techniques.

There are online services that support passphrase generation, such as `https://www.useapassphrase.com/` and `https://untroubled.org/pwgen/ppgen.cgi`.

When you use `https://www.useapassphrase.com/`, it generates passwords based on the number of words:

Figure 7.16 – useapassphrase has generated a four-word passphrase

You can select the number of words you need to create the passphrase and our given example has used a four-word passphrase. The advantage of passphrases is that they are easy to remember and extremely hard to guess or break. If you noticed, the approximate crack time is given by centuries.

When you use `https://untroubled.org/pwgen/ppgen.cgi` to generate a passphrase, a number of options are given, including the number of words, the length of the words, enabling random capitalization, and inserting digits. This page only generates data and will not be stored by any means and it shows the number of combinations possible:

Figure 7.17 – untroubled.org password generator can generate passphrases

The `untroubled.org` password generator can generate passphrases that are extremely hard to guess and break using common password-cracking tools. The source code for the password generator is available to download to confirm that this site does not collect this data, but the drawback of this passphrase is that it is not easy to remember. Users need to use password managers such as KeePass or BitWarden, which we discussed in previous chapters.

- **Using secure email service** – Using Gmail is secure but as we discussed, it does not provide the expected privacy. Google provides a complete ecosystem with many free services, which many users are fascinated by, but when it comes to privacy, Google keeps all our information and the traces of our activities, which is not our objective when it comes to privacy and anonymity. To maintain cyber anonymity, we need to use a more secure option such as ProtonMail. Being a Switzerland-based company, ProtonMail provides a higher level of privacy, apart from end-to-end encryption. Since Switzerland is not a member of the surveillance alliance **the Five Eyes (FVEY)**, an intelligence alliance comprising Australia, Canada, New Zealand, the United Kingdom, and the United States, the government cannot request customer data. Being an open source product, ProtonMail provides a secure and free email service that is accessible over a browser, Android, and iOS. When you compose a mail using ProtonMail, the email will be encrypted using a client-side encryption mechanism even before the mail is sent to the ProtonMail mail servers. ProtonMail provides customers with complete privacy – in other words, the user can control who has access to the user data, and not even internal employees of ProtonMail will have access to your emails.

- **User secure storage** – We all are fascinated by Google services, including Google Drive, but when it comes to privacy, Google Drive is no different. We store our data in the Google services and Google collects all our information and has access to all our data. If you want to maintain cyber anonymity, it's high time to think about secure storage options other than Google Drive. There are better and more secure solutions for storing your data. There are a few options when it comes to secure cloud storages that provide complete encryption:

  - `sync.com` – Provides client-side, end-to-end encryption, which will not even allow hackers or internal employees to access your data. Sync does not collect, share, or sell user information to third parties. The user can completely control their privacy when using Sync.

  - `pcloud.com` – pCloud uses TLS/SSL encryption and client-side encryption, which protects files even before they leave the client's device. Unencrypted files never leave the customer's device. They provide zero-knowledge privacy, which means encrypted keys will never be uploaded to the servers. They use 4096-bit RSA and 256-bit AES for encryption.

  - `Icedrive.com` – Icedrive provides client-side encryption to everything, including file and folder names. Everything will be encrypted even before it reaches the Icedrive secure cloud. Icedrive uses TwoFish encryption and provides zero-knowledge encryption so that only the user can access and view the data.

- **Use disposable email addresses when necessary** – If you want to receive an activation link, download link, or **One-Time Password (OTP)** only once, use a temporary disposable email address. Attackers use this method to harvest email addresses for future attacks in particular. They will come up with creative ideas such as registering to download an e-book or entering your email address to receive a download link. Once you enter your email address, you will start receiving a lot of unwanted emails and spam. To avoid your email address being harvested, you can use a disposable email address.

- **Stop overexposure** – We discussed many layers that collect your information without your knowledge or consent. Now, we know the best practices that will prevent your information from being collected from you, but still, you might be overexposing yourself. We discussed how an image can carry your information. This can happen on your social media, forums, emails, vacancy advertisements that you are posting, in telephone conversations with unknown parties, in telephone conversations with known parties in public areas, when filling out a form, and through various other activities without thinking about it. All these acts can overexpose you.

- **Don't share sensitive or personal information** – While interacting online, make sure you never share any personally identifiable information online including your address, phone number, and social security number. I have seen many people disclose their phone numbers in public when cashiers ask about a loyalty card. There are many loyalty programs, including in supermarkets, where when you are in the queue, the cashier asks what your loyalty number is. When you share your number, others in the same queue can memorize, write down, type, or record it on their phones.

- **Personnel assistance programs** – Personnel voice assistance systems such as Amazon Alexa make your life convenient and easy but when it comes to privacy, voice assistant programs are so bad. According to the Times Magazine, Amazon.com, Inc recruits thousands of people around the world to improve the Alexa digital assistant. These teams listen to the recordings captured by Alexa personnel assistance devices in homes and offices. Then, the recordings will be transcribed, annotated, and again fed into the software to eliminate the gaps. This shows that personal assistant programs are not that great when it comes to privacy. If you want to maintain cyber anonymity, keep personal assistant programs away.

- **Secure connectivity** – Whenever you interact with the internet, make sure the web applications are using `https://` URLs. Simply check your browser shows the padlock sign and secure connectivity. If you are communicating with any website which does not have `https`, there are chances that attackers can intercept your communication as the communication is not encrypted. **Hypertext Transfer Protocol Secure (HTTPS)** encrypts the communication between your browser and the web application using SSL encryption. You can check this by clicking on the padlock sign on the URL:

Figure 7.18 – Secure connection established by https

If the web application doesn't show the padlock sign as shown in the figure, your communication is not encrypted, which means that attackers can see what you are doing, even sensitive information such as your passwords and bank information, or whatever information you are communicating.

Developing a cyber anonymity mindset

When interacting with the cyber world, there are multiple components working together to establish connectivity. Once connectivity is established, data will flow through multiple layers. When you work on applications, data will be sent over multiple layers during client-server communication. As we discussed in the previous sections, we need to look at all these elements to maintain anonymity in the cyber world. Due to the vast number of elements involved, it's not easy to maintain cyber anonymity, as all these elements are collecting information as per the design. Especially for your typical user, concentrating on all areas is going to be hard. When you concentrated on the *Best practices to maintain cyber anonymity* section, you may have noticed that we must be extremely mindful to maintain cyber anonymity. The best way to maintain cyber anonymity is to develop a cyber anonymity mindset.

This can be started now and applied to devices that you use all the time – these can be mobile devices, laptops, desktops, smart devices, or even personal assistance devices. Just think about whether your own device is secure. You can assess whether your own device is secure by answering the following questions:

- Is my device physically secured? Can anyone steal my device? Do I leave my device unattended?

- Is my device protected with a passcode or biometric authentication?

- Is my device encrypted? If someone steals my device, can they access my data?

- Are the applications installed in my device known applications and installed by me? Or are there any applications that came with the device I'm not aware of?

- Is my device purchased brand new as a sealed pack or did I get it second-hand? Was it gifted?

- When I got the device, did I perform a factory reset before using it? This applies to a pre-owned or gifted device.

- Is my phone connection under my name? Do I have total control over the ownership of the mobile connection? Am I the one who is receiving my bills? Am I very sure that no one can access my billing information or divert ownership of my mobile number?

- Do I receive my phone bills and statements, physically or digitally? Does anyone else have access to this information? If I receive them physically, do I shred or discard them?

- Are the apps on my device installed from an app store (if it is an Android device, from Google Play Store, if it is an iOS device, from the App Store, and if it is a Microsoft device, from Microsoft Store)? Are apps given the minimum permissions?

- Do I connect to public, open, or untrusted Wi-Fi networks?

- Do I avoid clicking on links or opening attachments sent by unknown senders?

- Are the accounts connected to my device (Google Account, Apple ID, or Microsoft Account) secure and MFA-enabled?

- Are the secret questions provided to these accounts not guessable, even by my closest contacts?

If the answers to these questions are *yes*, to at least most of the questions, you will know you are using a secure device. If not, you can take action to make your devices more secure. With these questions, you can understand the importance of other elements connected to your devices, such as the accounts connected and billing accounts. For example, if someone can claim and get a SIM card for your number by claiming ownership or as a replacement for a misplaced SIM, they will have access to everything of yours, meaning they can receive all your OTPs (sent to reset your passwords and access devices and accounts including Google or Apple IDs). Then, you need to think about the other connections that are used to access the internet from your device.

This way, you can start developing a mindset to maintain anonymity in the cyber world. Once you started practicing, it will become a habit, and you will be suspicious whenever you come across anything that can potentially compromise your privacy and you will be able to prevent yourself from being exposed when on the internet.

Summary

This chapter focused on understanding cyber anonymity and the layers of cyber anonymity. We discussed the basics of cyber anonymity in the *Definition of cyber anonymity* section. Then, we tried to understand the difference between privacy and anonymity. Then, we defined and understood the different layers of cyber anonymity in the *Levels of cyber anonymity* section. We also discussed the best practices to maintain cyber anonymity and developing a mindset to maintain anonymity in the cyber world. During this chapter, we developed a set of skills for maintaining anonymity:

- Understanding cyber anonymity
- Understanding the difference between privacy and anonymity
- Understanding the different layers of cyber anonymity
- Best practices to maintain cyber anonymity
- Developing the correct mindset

In the next chapter, you will be provided with information on how to plan for cyber anonymity and the prerequisites to maintaining cyber anonymity. In the next part, we will try to understand the scope of access and the plan for connectivity and understand the level of access. Then, we will prepare the device and the applications for anonymity.

8

Understanding the Scope of Access

In the previous chapter, we discussed the levels of cyber anonymity, as well as how to develop a mindset and best practices to maintain cyber anonymity. This chapter provides information on how to set up cyber anonymity and requirements to maintain cyber anonymity under the following topics:

- Understanding the scope of access
- Planning for connectivity
- Understanding the level of access
- Preparing the device
- Preparing applications for anonymity

Understanding the scope of access

In the cyber world, the most used way of providing access to services is through an identity system. We discussed different types of identity systems in previous chapters. There are centralized and decentralized identity systems. Some identity systems are application based and some identity systems are web based. However, all identity systems keep a bunch of attributes that relate to identity. There are many web-based identity providers today that support **Single Sign-On (SSO)** – when configured, authentication will take place on the identity provider's identity system, which provides access to other services, such as service providers. Facebook, Google, and Microsoft are popular identity providers that can be integrated into other applications for authentication purposes.

When the authentication is completed, the next step is authorization. Authorization is maintaining a level of access. For example, if you take a company with a large number of employees, all the employees may have a username and password. That doesn't mean that everyone can access everything once they have access to the company systems. The level of access is defined by authorization, meaning different users will have different levels of access based on their job roles. Some users will be given more access and some users will be given less access to the system. For example, senior managers will be given more access and junior executives will be given less access to resources. This mechanism is often known as a **Least-Permission Policy (LPP)**. According to LPP, a user or the subject must be given the lowest possible permission to perform their job – no more than that. To provide LPP-based access, first systems administrators need to analyze the requirements.

Providing the required level of access is often complex. Most companies practice a principle known as **Separation of Duties (SoD)** to achieve this goal. The meaning of SoD is: no user should be given enough privileges to misuse the system on their own. LPP and SoD sound the same but their objectives are different. The objective of SoD is to distribute and divide important duties among different members of the team, so one person will not be able to commit any fraudulent activities that can cause damage to the system on their own. The objective of LLP is, even if duties and tasks were assigned, permissions are only given to perform the duty, no more than that. This prevents misusing permissions given to users.

When accessing any application as a user, the two concepts that are important to understand are how the application maintains LPP and SoD. Basically, these concepts are used by administrators to provide access and effectively maintain access. Sometimes we use built-in identity systems to authenticate, meaning the system will be creating login credentials for us. This typically goes through a registration process. Once registered, we receive an activation link to the registered email address. Once we click on the activation link, we will be given access to the system. Typically, this will be user-level access to the application. In order to get user-level access, we have to go through a lengthy registration process.

But we can use another option that most applications today provide, of using a readily available identity from another popular identity provider, such as one of the ones we discussed before. This option will redirect us to authenticate on their web application; then, once authenticated by their web application, we will be redirected to the service provider's application. Both mechanisms used identity to authenticate users to services. As we discussed before, access control is not a simple thing. It's very complex based on the requirements. Whenever we access any type of application, remember the way it provides access is based on identity. To create or use an identity, applications collect information from the user.

From a cyber anonymity perspective, we need to concentrate on the level of access that we need to maintain. If we need more access to the application or system, the authentication provider would be collecting more information from us. Authentication and authorization are based on the identity that we need to create in order to access the resources. In other words, when we must authenticate, we need to create an identity by providing our information. Then, the system will keep our information, which we do not need in this case. Systems that do not need to authenticate will not collect information directly. These systems will collect information using indirect methods, such as third-party cookies, which we discussed in the previous chapter.

The next question would be that even if the systems or resources we access need an identity for authentication purposes, do we need to use authentic or original information to create an account? For example, can we use a disposable email address with a pseudonym to create identities to access resources? That depends on the requirements. For example, a banking app, or any other app that requires authenticating real users, requires you to provide authentic information; otherwise, it will not serve its purpose. But if you need to access certain services, let's say a resource that provides e-books or a news feed, you can use a pseudonym or temporary email address to create an account to gain access and access the resource without sharing your personal and sensitive information.

Planning for connectivity

When you are surfing the internet or accessing any service over the internet, connectivity plays a major role. First, your device should be connected to the internet in order to access any service. When the device is connected, you will get access to the desired service. But the real concern is, your internet provider collects and shares your information with different service providers. Previously, we discussed the importance of not connecting a device to open and public networks. When you connect a device to an untrusted network, you have no idea who has access to the network and how much data they have access to. Unencrypted network protocols are especially prone to this kind of attack and information disclosure. If you need to maintain cyber anonymity, it is better to use a trustworthy internet connection. There are different types of internet connections available today.

Types of connections

Even though there are a lot of connections that provide internet access, there are differences and some concerns. Not only individuals but also many businesses and companies use different types of connections.

Dial-up connections

This was the only option before introducing broadband connections, but now this type of connectivity is very rare. But still, some service providers offer this connection type in rural areas where other connections are not available. This is the slowest connection type; technically, the connection is established by the modem connected or built into a user's computer dialing the service provider's network and establishing the connection. When the connection is established, users can surf the internet very slowly, provided through **Public Switching Telephone Network** (**PSTN**) – the network used for regular telephone lines. Most **Internet Service Providers** (**ISPs**) have stopped providing this connectivity.

When it comes to security, it's hard for someone to intercept traffic generated from your system to the ISP. Mostly, a dial-up connection establishes the connection from an individual device to the ISP network. Typically, your device is not connecting to a local network, but you will get assigned a dynamic IP once the connection is established. Once you have disconnected from the ISP and established the connection again, you will be assigned a different IP address. Typically, a dial-up connection is safer as interceptors have comparatively lower opportunities when using dial-up connections.

Broadband connections

This term refers to any internet connection that provides a wide-bandwidth data transmission and supports multiple signals and ranges of frequencies. Typically, broadband connections don't charge customers for the connection time. There is a range of broadband connection types available today in many countries and service providers.

Cable connections

Mainly, cable connections are used by companies and apartments using television cables, also known as **coaxial cables**. Through cable connections, users are connected to the internet with higher speeds with stable connectivity. Cable connections have low physical security and can be easily intercepted without interrupting the transmission and without being detected.

Leased lines

Typically, companies use leased lines with faster and more reliable connections over dedicated links. Generally, this provides a connection between the ISP and the company. The main advantage of a leased line is the speed is not shared with anyone. The total bandwidth for the company is determined by the ISP. This provides very fast and reliable links for the company's operations. When it comes to security, a leased line provides a high level of security as the connection is dedicated and no one else is using the same link.

DSL connections

Digital Subscriber Line (**DSL**) connections are used by small and medium organizations and provide high-speed internet over telephone networks. Depending on the area and the service provider, the connectivity speed can differ. Providing a DSL connection is easy as it uses typical telephone networks and usually most small and medium organizations already have a telephony connection. DSL uses shared bandwidth; hence, the speed can be different depending on the service provider or the time of the day. When more users are connected to the DSL service provider, the speed can go down. DSL connectivity provides different levels of security risks as computers and devices are connected all the time, so bad guys can try to compromise systems whenever they want.

Fiber connections

Fiber connections are very popular among domestic users as well as businesses due to their speed and reliability. Fiber connections use light as the medium of data transfer. Even if the ISPs use fiber to provide a reliable connection, they still use DSL or cable to deliver the connection to customers. Even for domestic use, fiber provides very fast connectivity and the connection can easily be intercepted as it uses light as the transmission medium.

Ethernet

Once the internet connection is provided, if you have multiple devices to be connected to the network, this is the preferred method in many companies. Typically, a **Local Area Network** (**LAN**) is built using Ethernet. Ethernet cables are used to connect devices. When it comes to security, an Ethernet network provides a higher level of security when compared to Wi-Fi. To access an Ethernet network, your device must have physical access to the Ethernet network, whereas a Wi-Fi network can be connected to without physical access to the network.

Wi-Fi connection

Wireless fidelity, commonly known as wireless networks, is commonly used in houses, restaurants, hotels, and even companies as Wi-Fi connectivity provides mobility and flexibility. Users are given the flexibility to connect from anywhere within the signal range. The medium of connectivity is air – radio frequency to be specific. Mostly, Wi-Fi is the preferred method of connectivity for faster and more reliable connections, such as cable, DSL, leased lines, and fiber links, to the endpoints. Even though Wi-Fi provides flexibility and mobility, it introduces a range of newer security risks. Since Wi-Fi networks provide connectivity through the air, the possibility of interception is high. Unlike Ethernet networks, physical connectivity is not required for attackers to launch attacks on Wi-Fi networks.

Mobile broadband

Many mobile ISPs offer high-speed broadband internet over **Third-Generation** (**3G**) wireless and **Fourth-Generation** (**4G**) wireless networks. Using mobile networks, ISPs provide internet using 3G and 4G technology. 4G technology provides faster internet connectivity than 3G networks. Recently, some service providers have started offering **Fifth-Generation** (**5G**) wireless. Mobile broadband is typically more secure than other network types we discussed before as when you use 4G or 5G networks, the data being shared is encrypted. When you use a Wi-Fi network, the connection is typically encrypted using symmetric encryption. But users who are connected to the same network can still see others' communication. When you use 4G connectivity, you are using a secure connection and the communication is encrypted. In other words, a 4G connection is much safer than a public Wi-Fi network. On the other hand, 4G networks are much harder to crack than Wi-Fi networks. If you want to ensure secure communication, it is better to use 4G mobile data than an untrusted Wi-Fi network as 4G communication is encrypted by a cellular network using a 128-bit key.

Considering the different connectivity types we discussed, some connectivity types provide more security than others, such as 4G mobile broadband connections. Some connectivity types offer faster access to the networks, such as fiber and leased lines. The newly introduced 5G is more secure than 4G.

Also, we discussed security considerations in different types of connections. As we discussed before, connectivity should be trustworthy if we need to maintain ethical cyber anonymity as attackers can sniff the network and compromise confidentiality otherwise. When the appropriate connection is selected, we need to ensure the connection is secure. We can't implement any security for some connections as the security is entirely controlled by the service provider. However, when the connectivity is decided and connected to the local network, we need to ensure the local connectivity is secure and safe. Let's see how we can make the network safe and secure to ensure confidentiality.

How to secure a home network

Most of our home networks today are powered by Wi-Fi routers. The router is connected to the internet using different kinds of connections, as we discussed previously, including fiber, cable, 4G, or DSL. Mostly, the whole household uses the same Wi-Fi to connect different types of devices to the internet. Sometimes, visitors or guests can even ask for the Wi-Fi password to connect to the internet. This means many people have your Wi-Fi password saved in their devices, maybe when they passed by, so they can connect to your Wi-Fi router without even your knowledge. As we discussed before, unlike Ethernet connections, Wi-Fi does not need physical connectivity or cable. If they are in the signal proximity, they can connect to it. There will probably be buildings, such as apartments or restaurants, around your house that receive the signals from your Wi-Fi router.

Sometimes, I've seen people going around with mobile devices to check what the maximum distance to receive signals from their Wi-Fi network is. But this is not a very accurate method as attackers can receive Wi-Fi signals from a longer distance than you think using high-gain antennas.

In 2015, at the DEFCON security conference, security researchers planned to introduce a hardware device known as **ProxyHam**, which could connect to public Wi-Fi networks from 2.5 miles away. While it was not introduced as anticipated at the 2015 conference, the idea and technology was made available to the public. ProxyHam is a dictionary-sized device that uses 900-megahertz radio signals to connect to a device with a Wi-Fi antenna or a public Wi-Fi network from 2.5 miles away. In other words, if someone plants a Wi-Fi dongle on a public device, it can be used as an entry point to connect to the network from 2.5 miles away to maintain anonymity. This proves that attackers can receive signals from much longer distances than we think as usually, we think Wi-Fi signals are only accessible within around a 100-foot diameter.

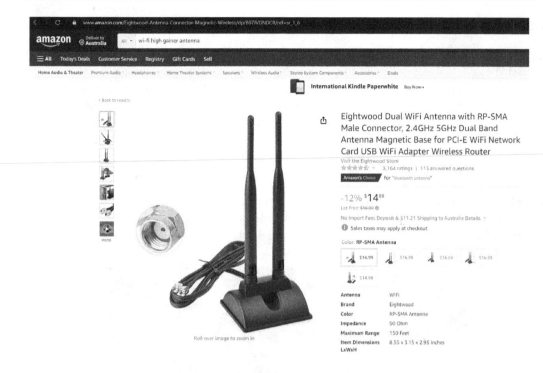

Figure 8.1: High-gain antennas to receive signals from longer distances

This Wi-Fi high-gain antenna can be used to boost signals or receive signals for long distances. Attackers can employ these types of high-gain antennas to compromise your networks. Since Wi-Fi communication is based on transmitter and receiver communication via radio frequency, we cannot be certain that only the people close to the network will be receiving signals.

As we discussed, unlike Ethernet networks, your Wi-Fi signal goes beyond your premises and your walls. Sometimes, some areas in your house might not receive signals but attackers can still connect to your Wi-Fi from a greater distance if they use high-gain antennas. When we share our Wi-Fi passwords with many visitors and guests, we will never know how many people or devices are connecting to our Wi-Fi network. Considering these facts, one of the main things to do is restrict access to your Wi-Fi network and control who can access and use your Wi-Fi signals. The next issue is that of the signal footprint; as we discussed in *Chapter 6, Artifacts that Attackers Can Collect from You*, anyone who receives the signals can capture data on your network. They will not be able to access data as the communication is encrypted, but if the Wi-Fi password is not complex, they can even crack the key and access the content of communications, including your personal and sensitive information. Let's see how we can secure a Wi-Fi network.

Configuring a strong Wi-Fi password

Configuring a strong password is a must. Typically, this must be over 12 characters as attackers can try to crack the password using readily available word lists. We can also use passphrase generators, as we discussed in the previous chapter, to generate a passphrase for the Wi-Fi router. This can be done by accessing your Wi-Fi router control panel. Typically, when you connect the router for the first time, you can use the router's default password printed on the router to connect to the router. Once connected, the router will assign IP addresses to your device. Modern routers are plug-and-play devices; users can just plug in the router and it starts working without additional configuration. But typically, routers use very basic Wi-Fi passwords that can be cracked by attackers even before you connect.

Once the device is connected to the Wi-Fi network using the default password, you can access the device's IP configuration. You can open the terminal and enter `ipconfig`, if the device is Windows, or enter `ifconfig` if the device is Linux. This will display the IP configuration for the device, obtained from the Wi-Fi router.

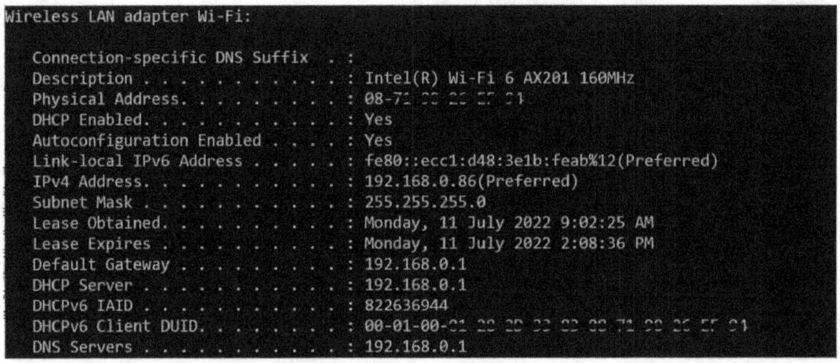

```
Wireless LAN adapter Wi-Fi:

   Connection-specific DNS Suffix  . :
   Description . . . . . . . . . . . : Intel(R) Wi-Fi 6 AX201 160MHz
   Physical Address. . . . . . . . . : 08-7? ?? ?? ?? ?1
   DHCP Enabled. . . . . . . . . . . : Yes
   Autoconfiguration Enabled . . . . : Yes
   Link-local IPv6 Address . . . . . : fe80::ecc1:d48:3e1b:feab%12(Preferred)
   IPv4 Address. . . . . . . . . . . : 192.168.0.86(Preferred)
   Subnet Mask . . . . . . . . . . . : 255.255.255.0
   Lease Obtained. . . . . . . . . . : Monday, 11 July 2022 9:02:25 AM
   Lease Expires . . . . . . . . . . : Monday, 11 July 2022 2:08:36 PM
   Default Gateway . . . . . . . . . : 192.168.0.1
   DHCP Server . . . . . . . . . . . : 192.168.0.1
   DHCPv6 IAID . . . . . . . . . . . : 822636944
   DHCPv6 Client DUID. . . . . . . . : 00-01-00-?? ?? ?? ?? ?? ?? ?? ?? ?? ?? ?1
   DNS Servers . . . . . . . . . . . : 192.168.0.1
```

Figure 8.2: ipconfig /all command shows the IP configuration of the device

Once you retrieve the IP configuration, typically, the default gateway shows the IP address of the Wi-Fi router. Generally, all Wi-Fi routers today have a built-in web server that hosts a web application that can be used to configure the device. The best way to access it is to enter this IP in the browser. For this example, we need to enter `http://192.168.0.1`. When you enter this in the browser, it will take you to the authentication window. Typically, the credentials are printed on the Wi-Fi device or user manuals. Alternatively, you can Google default passwords to access the device.

Then, you need to navigate to the Wi-Fi security options and change the Wi-Fi password on the devices. We need to configure WPA2 as the security mode. If your device supports WPA3, that's the best mode now, but it may not be available on your device as WPA3 was introduced recently.

Then, you can change the password to a stronger passphrase to protect it from password-based attacks. This includes changing both the security mode and the password. Previous WEP encryption is easy to break.

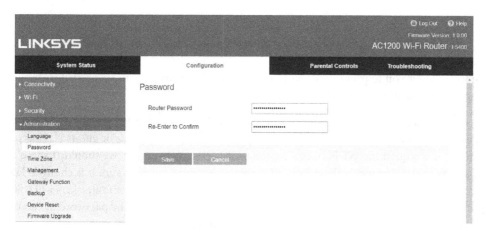

Figure 8.3: Configuring a complex passphrase to a Wi-Fi password

Once the proper passphrase and security mode are configured, we must change the device password. This is essential as any person who has access to the network can easily access this and even view the Wi-Fi password in plain text, as you can see in the preceding screenshot. Protecting the router password is equally important.

Figure 8.4: Changing the device password

You can configure the device password to protect the device. Also, upgrading the device's firmware is important as older firmware might have vulnerabilities that could lead attackers to compromise the Wi-Fi device. Some devices may even lead to downloading **Read-Only Memory** (**ROM**) and accessing all configuration settings on the device, including configuration data and credentials to the device. When upgrading the firmware, make sure to download new firmware from the original website of the vendor as attackers can distribute malicious firmware, which can lead to attackers planting a backdoor to the device or bricking the device. (Bricking refers to making a device permanently unusable.)

Some Wi-Fi routers provide a feature known as **guest access**, which provides access to the network temporarily without using the Wi-Fi password. If a device supports this function, you can create the guest network on the same device, and it will allow guests to access the internet but with limited accessibility to the network. You can later change the guest password.

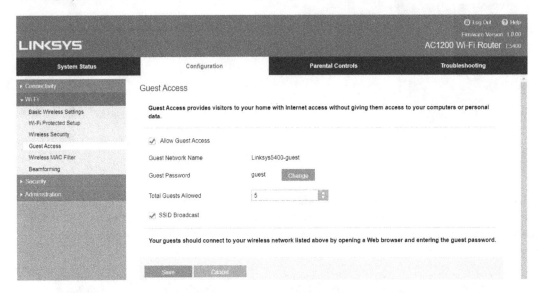

Figure 8.5: Enabling guest access to provide temporary access to the Wi-Fi network

Guest access allows you to provide access to the internet for temporary users or visitors without exposing your real network credentials.

We also need to make sure we limit access to who we share the password to our private Wi-Fi network with; if we share it with everyone, it may as well be public. Sometimes, friends or guests visiting your house might also request the Wi-Fi password; unless it is extremely urgent, we shouldn't share our Wi-Fi password with anyone. If the password is shared, we need to make sure it is changed later. Passwords must be changed frequently to prevent different types of attacks. Some attackers will collect data packets to break the password later. Depending on the complexity of the password, it can take some time to break the password. If you have a practice of frequently changing the password, even when attackers have compromised the password, they will not be able to access the resources as the password is changed when they break the old password.

Another approach to improving the security of a Wi-Fi network is disabling the SSID broadcast. **SSID** stands for **service set identifier**, which is commonly known as the **network name**. As you can see in *Figure 8.5*, there is an option to enable or disable the SSID broadcast. When you enable the SSID broadcast, devices will be able to see that there is a Wi-Fi network. When you disable the SSID broadcast, strangers will not be able to see that there is a Wi-Fi network. This can increase security by not showing the Wi-Fi network when attackers try to scan the network. Even if you disable SSID broadcasting, there are tools that attackers can use to view these networks, but still, you can be protected against general scanning.

Device configurations to make it secure

There are various other settings that can make your device secure. These settings are entirely device specific; some devices have these settings and some may not offer them:

- **Wi-Fi Protected Setup (WPS) configuration**: WPS is a feature available in many Wi-Fi routers that allows users to easily connect the device to the network. Typically, WPS offers two methods of connectivity. The first method is using the WPS button on the Wi-Fi device. Users can press the button while selecting the SSID on their device, which will connect the device to the Wi-Fi network without having to provide a password. If you configure this method, the router must be physically secured as anyone who can access the Wi-Fi device physically can connect to the network.

  The second method is using an eight-digit numeric code generated by the device or configured by the user. With today's computation power, the eight-digit numeric code can be cracked very easily. If the device has a WPS feature, turning off this feature can increase security.

  If the device supports MAC filtering, it's a great idea to configure this. This will restrict only devices with permitted MAC addresses to access the internet. When it comes to user-friendliness, MAC filtering is not preferred, but if the consideration is security, enable MAC filtering to provide safer access to the internet as only devices with permitted MAC addresses will be allowed.

- **Enabling firewalls**: If the device has built-in firewalls, you can enable them. This will prevent unwanted traffic from coming in, including anonymous internet requests. A built-in web filter can help us to prevent information collection using Java, ActiveX, and cookies.

Figure 8.6: Enabling a firewall on a Wi-Fi device

Understanding the level of access

Level of access is typically defined as authorization, which we discussed in the first section of this chapter. When it comes to cyber anonymity, we need to understand the level of access required based on our requirements. If you need a higher level of access, typically, you need to go through an authentication process. If you need a lower level of access to the resources, you might not need any type of authentication. In the first section of this chapter, we discussed what authentication and authorization are. Also, we discussed applications such as banking applications, which need to verify the user's authenticity before providing access. Some services, such as news websites, wouldn't require any authentication. Some applications require authenticating but not all applications would not need to authenticate you.

It's important to understand what level of access we need from the application or the services. Then, we can decide whether we need to create an identity. Some applications and resources might collect information from you even if it is not required. In these scenarios, we can refrain from sharing our sensitive information or personal information. Let's try to understand the different levels of access defined in various scenarios:

- **Administrators of superuser access level**: This is the highest level of access given to any device, application, or resource. Anyone who has this level of access can do pretty much anything they want, including making modifications and reconfigurations. Every permission, including read permission, is applied to these users. Typically, users are not given this level of permission. You will be given this permission if you are the owner or administrator of the resource. Hackers and attackers are dying to get this level of access to any resource.

- **Protected items access level**: This level of access is given for items or specific resources. Previously, we discussed SoD. In this case, separate permission will be given to different subjects. In this case, specific permission will be given to specific items or resources. Users with this permission will not have modify permissions but they can perform every other operation. For example, a user with this permission will be able to perform anything within the scope specified, but they will not be able to assign another user to have this level of permission. This means Permission for modification is not allowed.

- **Registered users**: Any user who has credentials or a valid username and password combination is categorized into this category. They can just access the resource, modify only specific settings, and mostly make changes within their own account. Some resources allow registered users to maintain their profile within the resource, so they can update their user information, including name, address, contact information, and even credit or debit card information within the profile.

- **Unregistered or anonymous users**: This type of user will always have limited access. Usually, only low-sensitivity information will be available to these users. Organizations use other techniques to collect information from these users, including cookies.

Once we understand the level of access, we know what we can do within the resource. Mostly, we will be given access as registered users or unregistered users. Depending on the requirements, we can decide whether we want to register with the resource. If registration is not required for our requirements, maintaining an unregistered status is best as it does not need our information. If registration is required, we need to maintain a minimum set of information as in today's world, even web applications can do a lot. If you access `https://whatwebcando.today`, you will be amazed by the features of web applications.

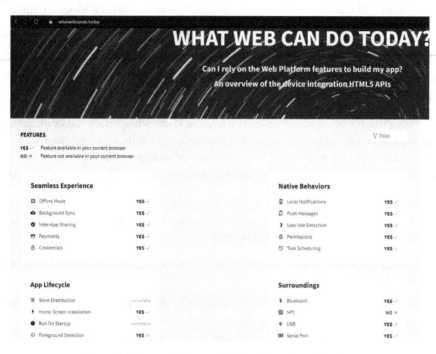

Figure 8.7: Some of the capabilities of a web application

Even when you access as an unregistered user, a web application can collect information about you. We discussed how a browser can collect information using cookies. Similarly, web applications accessed over the browser can collect information about your geolocation, operating system, input touch gestures, device battery status, device memory, recording media, advanced camera control, audio and video capturing, Bluetooth support, USB support, and **Near-Field Communication** (**NFC**) support.

With all these capabilities today, it's not easy to stay anonymous while surfing or interacting with the internet. As we discussed before, when you access the internet using browsers or applications, you will be tracked. All the layers that we discussed are watching and collecting information about you. Even operating systems collect information about you. Especially, if you are using the Windows operating system, you will be traced using telemetry data. Venders specify that they use telemetry data to improve the software and improve the user experience. However, if your motive is to stay anonymous, this is something you don't want to happen and there are better options for this. The Linux operating system, in particular, offers a range of distributions that support being anonymous online.

It's common for attackers, hackers, and cyber terrorists to maintain anonymity as they don't want to be traced and prosecuted. They use multiple technologies to prevent being traced. To achieve this, they maintain anonymity in various layers. But hackers are not the only people who wanted to become anonymous. There are many ethical reasons to be anonymous in today's world. As an example, some people want to share their thoughts without being victimized on various forums. They want to maintain anonymity so no one will identify and discriminate against them due to their ideas and thoughts.

In some countries, people may want to voice their opposition to the government but are scared of the government's retaliation; thus, if they express their thoughts, they would want to be anonymous.

Also, for privacy and security concerns, users might want to maintain anonymity as they don't want to be exposed to hackers or to the public.

Whatever the objective for being anonymous, you need to know who you should be anonymous in front of, or should you retain anonymity in front of everyone? As we discussed previously, anonymity protects your identity, not your actions. There are many advantages to being anonymous as your actions are visible but others won't know your identity. Let's see what other main advantages of being anonymous are:

- **Ensuring freedom of speech**: Once anonymity is guaranteed, you can share your real thoughts without being judged or discriminated against.

- **Minimizing tracking**: Trackers will have limited chances to trace and monitor your behavior patterns on the internet.

- **Information security**: Once you maintain anonymity on the internet, it will make attackers harder to identify and launch attacks on you as your identity is not visible. This will make it much harder for attackers to compromise your systems and steal your information.

- **Personal security**: Since your identity isn't visible, no one knows who you are. This will increase your security.

Preparing the device

Sometimes, people think the internet was built on anonymity as no one is regulating the internet or no one can regulate the internet due to its architecture. But when you analyze the components that are part of communication, you will realize that the privacy of the internet is hardly assured. While it might not be governments or employers that are spying, larger players on the internet, such as Google, Facebook, Amazon, and Microsoft, collect heaps of information for different purposes. Even though we use our own devices, private internet connections, and selected web browsers, these companies collect information and use it for product improvements, introducing new products, understanding new trends, and studying your behavioral patterns on the internet.

Some time ago, almost all the devices on the market had proprietary operating systems, such as Windows, iOS (the proprietary operating system used on Apple mobile devices), Symbian (the operating system used on personal digital assistance devices), **Research in Motion** (**RIM**, the proprietary operating system on BlackBerry devices), and Danger OS (also known as Hiptop or Sidekick, a Java-based proprietary operating system used by T-Mobile). At this time, there were many hardware vendors worldwide that didn't have a proper operating system to run on their devices, especially for touchscreen devices. The Android operating system became a solution for this.

The Android operating system was developed for mobile devices by a consortium of developers, commonly known as **Open Handset Alliance**, which was commercially sponsored by Google. Android was developed based on a modified version of the Linux kernel and open source software. Even though most versions of Android are proprietary, Android has kept the source model open source. That has given the option to develop variants of the Android operating system on different types of electronic devices, including TVs, game consoles, audio systems, digital cameras, and wearables. This has made Android one of the most popular operating systems among many vendors and achieve higher market share. This means for many users, the first devices they used ran on the Android operating system. Most Android devices come with a bunch of additional proprietary software pre-installed. This proprietary software includes Google mobile services, Google Play, and core Google services.

This has given Google the opportunity to collect information about users on a range of electronic devices. Not only Android but other mobile operating systems and desktop operating systems also collect information from users. What these vendors say is, they want to provide you with a better experience while using the device. They also say that they collect information to improve the software, fix bugs, and introduce newer and better features. This is true; they use the collected information to provide a better user experience. However, they cannot deny the fact that they also use this information for their product improvement, to provide custom features and functions to end users. Some information they collect can be sensitive.

Other large companies aside from Google, such as Microsoft, Apple, and Amazon, do the same; they also collect user information for various requirements, as we discuss in this book. How to prepare a device to be anonymous is another useful area to look at. Under this topic, we will look at how we can prepare various devices to maintain anonymity by doing the following:

- Reducing the footprint on the operating system
- Reducing the temporary and permanent files created on the operating system
- Removing temporary and permanently created files after interacting with the internet
- Minimizing the device footprint on various resources we interact with on the web

Various operating systems act in different ways in the context of anonymity. Depending on the device you use, preparation can be varied. We will be discussing some of the secure operating systems that can be used to interact with the internet without creating a considerable footprint. Also, we will see how to configure commonly used operating systems for the previously mentioned requirements. If you are using an operating system such as Windows, it collects information via telemetry. Because of this, it is better to use the Linux operating system as there are plenty of Linux distros that offer features that enhance cyber anonymity. But in today's world, most apps and services we use for personal and professional requirements may run on the Windows operating system, so we cannot just stop using Windows due to privacy concerns. Virtual machines and live boot systems can be a lifesaver in this regard. Let's look at the benefits of using virtual machines and live booting systems.

Virtual machines

Virtual machines are the machines that run on top of your physical machines, typically on a virtualization layer. These machines can work like normal physical machines. A virtual machine utilizes the hardware resources of the physical machine. Ideally, you can run multiple virtual machines on a single physical machine if the physical machine has sufficient hardware. Typically, the physical machine is referred to as the host machine and the operating system of the physical machine is known as the host operating system. The operating system on the virtual machine is known as the "guest operating system." Depending on the host operating system, there is a range of options available for virtualization.

If you are using Windows as the host operating system, it has built-in virtualization technology for Windows known as Hyper-V. Hyper-V is available in client operating systems such as Windows 10 and server operating systems such as Windows Server 2012, Windows Server 2012 R2, Windows Server 2016, Windows Server 2019, and Windows Server 2022. You can build guest machines on Hyper-V. This gives us the option to install Linux distributions on the Windows 10 operating system. Once you use a virtual operating system, typically, it reduces the footprint, and if you want to you can even revert it to a previous snapshot, which removes every activity you performed on the virtual machine. Including Windows Hyper-V, there are many virtualization options available, as follows:

- **Hyper-V**: Hyper-V is a feature that comes with Windows operating systems without additional payment that supports a range of guest operating systems, including many versions of the Windows client and server operating systems (Windows XP, Vista, Windows 7, Windows 8, Windows 8.1, Windows 10, Windows Server 2003, and all other Windows server versions), Linux, and FreeBSD.

- **VirtualBox**: VirtualBox is an open source virtualization technology that can be used to build enterprise virtualization, which supports the creation and management of virtual machines. VirtualBox runs on various host operating systems, including Windows, Linux, and FreeBSD. VirtualBox supports transferring guest machines from one host to another. The best thing about VirtualBox is it's free!

- **VMware Workstation and VMware Workstation Player**: VMware Workstation is a reliable virtualization technology for professionals. VMware Workstation Player is a basic version and is free to use, whereas VMware Workstation Pro is a commercial and feature-rich version. Both the VMware Player and VMware Pro versions support similar host operating systems and hardware. VMware is an enterprise-level virtualization technology that is compatible with most 64-bit Windows and Linux host operating systems, including Windows 8, Windows 10, Windows Server 2012, and later server operating systems, Ubuntu, Red Hat Enterprise Linux, Oracle Linux, and CentOS.

- **VMware Fusion**: VMware Fusion is an offering from VMware that supports Mac operating systems that is ideal for application development and testing. VMware Fusion allows developers to run multiple applications on multiple operating systems simultaneously and integrate with many development tools. It is mostly used by cloud application developers.

- **QEMU**: QEMU, also known as B, is an open source tool written in the C language that supports a range of host operating systems, including Windows, Linux, and FreeBSD. This is a lightweight virtualization and emulation platform that supports full system emulation.

- **Citrix Hypervisor**: Citrix Hypervisor offers desktop virtualization that supports a range of host operating systems, including Windows 10. It provides simple management for testing intensive workloads, which allow users to enjoy enhanced graphic workloads. Citrix supports centralized virtualization management.

- **Red Hat Virtualization**: Red Hat Virtualization is an open source platform that offers virtualization for Linux and Windows operating systems. Red Hat Virtualization offers a range of features, including a single management and provisioning feature for new virtual machines, the ability to clone existing machines easily, and easy setup and management.

- **Kernel Virtual Machine (KVM)**: KVM is virtualization software for Linux that offers a virtualization infrastructure module. This is free software that supports Windows, Solaris, and Linux.

- **Mobile emulators**: If you would like to run the Android operating system virtually on your Windows 10 machine, you can consider a range of Android emulators, including BlueStacks, MEmu, Nox, GameLoop, Bliss OS, Xamarin, Phoenix OS, and Android Studio.

Depending on your expectations of being anonymous, you can use virtual machines as an option when interacting with the internet as it provides a secure and hassle-free method to maintain a very low footprint and anonymity. All the virtualization techniques that we discussed previously provide similar functionality for our requirements. I will explain how we can use VMware Workstation to create virtual machines to maintain anonymity. First, you can download and install VMware Workstation. Once you have installed it, it's important to configure the virtual network in a different network range and enable **Dynamic Host Control Protocol (DHCP)**, which can assign IP addresses automatically for our virtual guest operating systems. Then, you need to access **Virtual Network Editor** to configure the required configurations.

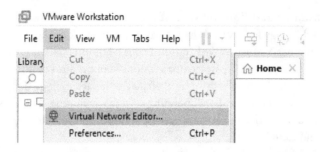

Figure 8.8: Select Virtual Network Editor... in VMware Workstation

When the **Virtual Network Editor** option appears, you need to select the **Change Settings** option to configure the required settings. Then, it will open the **Virtual Network Editor** configuration window.

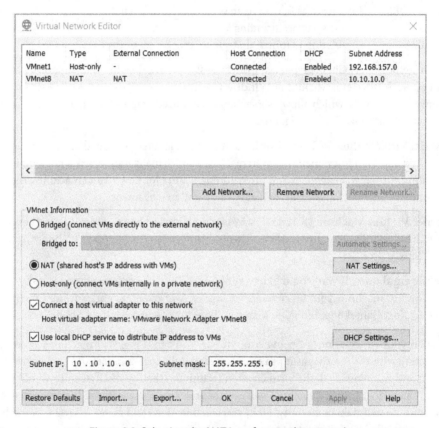

Figure 8.9: Selecting the NAT interface on the network

You need to configure the subnet IP and the subnet mask. As shown in the preceding figure, I used the 10.10.10.0 subnet IP and the 255.255.255.0 subnet mask for this demonstration.

The meaning of this is, all the virtual guest operating systems will have IPs from this IP range. Once you have configured the subnet IP and the subnet mask, you need to configure DHCP. Click on **DHCP Settings…** to do that.

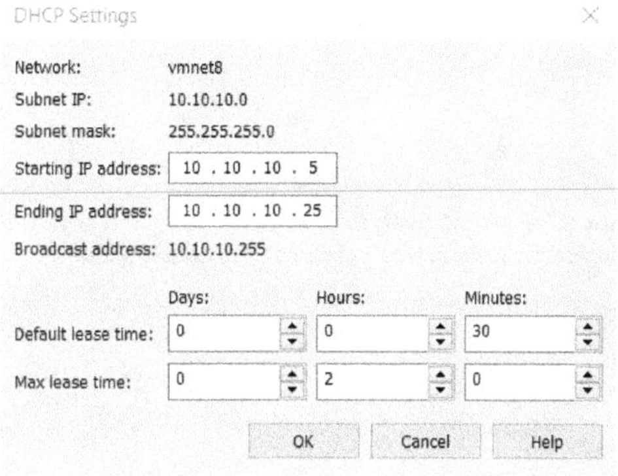

Figure 8.10: DHCP configuration for virtual machines

In the DHCP settings, you can configure the range of IPs that you want to assign to virtual machines. I configured `10.10.10.5` to `10.10.10.25` as I can assign 21 IPs to my virtual machines. You can configure any range you want, and then virtual machines will obtain IPs from this range. When you save the settings, it will be ready to assign IP addresses. Then, you need to configure the gateway address. You can do that by configuring the NAT settings.

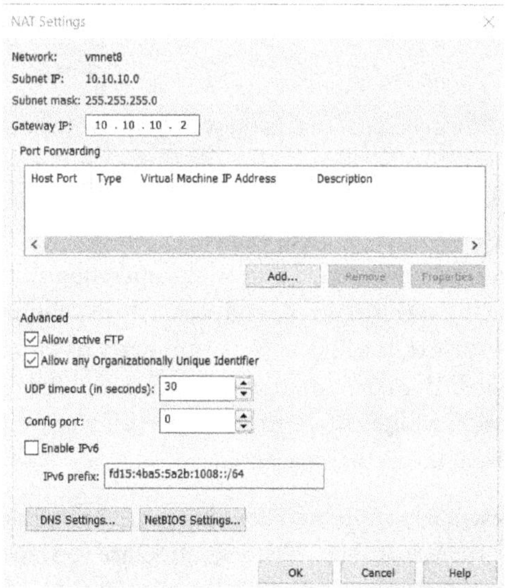

Figure 8.11: Configuring the default gateway in the NAT settings for virtual machines

Now, your virtual network is ready to host virtual machines. Let's see how to create a virtual machine and configure it to access the internet. You can create a virtual machine by clicking on the **New Virtual Machine** option under the **File** menu on VMware Workstation. You can select the **Typical** installation when building a virtual machine. On the guest operating system installation window, select the **I will install operating system later** option and continue. Then, you will need to select the guest operating system that you are going to install.

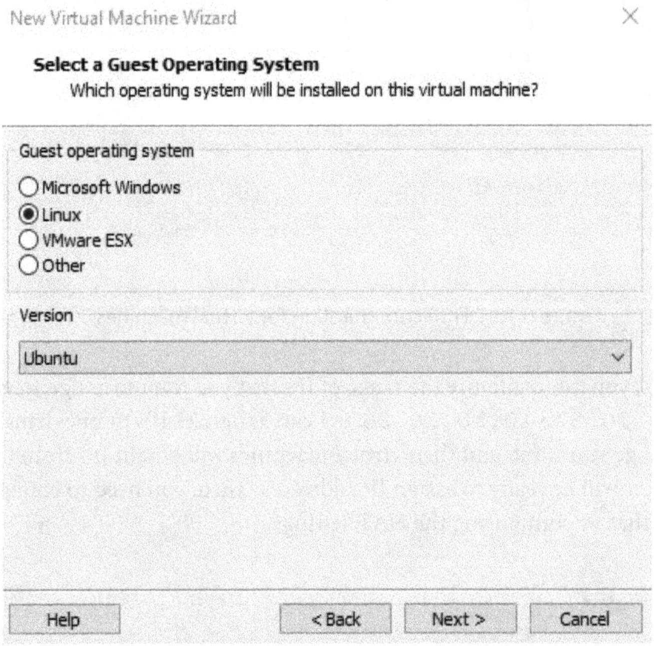

New Virtual Machine Wizard   ✕

Select a Guest Operating System
Which operating system will be installed on this virtual machine?

Guest operating system

○ Microsoft Windows
◉ Linux
○ VMware ESX
○ Other

Version

| Ubuntu   ⌄ |

| Help | | < Back | Next > | Cancel |

Figure 8.12: Selecting the operating system you need to install

Select **Linux** as the operating system and **Ubuntu** as the version. When you click **Next**, it will ask for a name for the virtual machine and the location to store data. Then, it will prompt you for the virtual hard disk size that you need to configure. Depending on the operating system, the disk size required can be different. Configure 40 GB for this example and select **store virtual disk as a single file** as it's easy to manage. When you click **Next**, it will give you a summary of the configuration and you can customize the hardware if needed. This configuration can vary depending on the hardware resources of the host machine as the host machine also requires hardware resources. Then, you can click on **Finish** to complete the creation of the virtual machine.

This method can be used to install any supported operating system on the virtual machine. We will discuss a few Linux operating systems that support anonymity. Once you have decided on the operating system to install on the virtual machine, you can simply download the ISO file of the operating system and install it on the virtual machine that you created.

Figure 8.13: Customizing the hardware if you require

Now you are ready to install the operating system on the virtual machine created. Since we have configured the virtual network, whatever operating system we installed will obtain the IP address automatically.

Using live boot versions

Using live boot versions is another way to keep our footprint to a minimum. In the past many operating systems supported live boot CDs and DVDs. Today, the usage of CDs and DVDs has become very limited. The idea of live boot versions is to run the operating system without installing it onto the device. You can run the operating system from the storage media and start using it without installing it. Since CDs and DVDs are read-only storage media, once the operating system is shut down, all temporary files and traces created will be removed from the system automatically. Nowadays, we can use USB live boot systems. Many Linux-based operating systems support USB live boot. The main advantage of USB live boot is it does not install anything on the system. You can plug the USB drive into the computer and start using it straight away. Kali Linux and Parrot Security are examples of live boot versions. These operating systems support live boot on virtual machines. For example, you can download a Kali Linux ISO file and run a live system on the virtual machine we created. The only configuration you need to do is download the ISO file and configure it on the installation media on the virtual machine settings.

Another option is to convert ISO to a USB bootable disk and use the USB bootable disk to boot the system. Not all operating systems support live boot. But you can convert ISO to a USB bootable disk by using `isotousb.exe`, which can be downloaded from `http://www.isotousb.com/`.

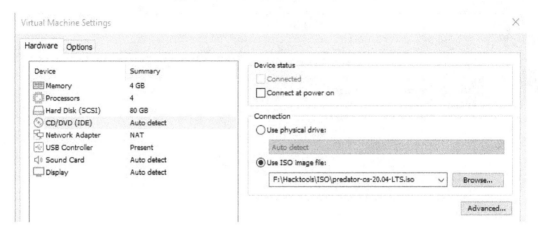

Figure 8.14: Booting a virtual machine using ISO

As the preceding figure shows, you need to go to the virtual machine settings, configure a connection to the ISO image file, and select the downloaded ISO file. Then, you need to save the settings and power on the virtual machine. The virtual machine will then start up with the operating system on the ISO file. If the operating system supports live boot, you can select live boot to boot the operating system. There is a range of Linux operating systems that support live USB boot:

- **Linux Lite**: This is a Ubuntu-based operating system that supports live boot with a range of pre-installed applications.

- **Peppermint OS**: Peppermint OS is a lightweight Linux distribution that supports live boot functionality with pre-installed applications.

- **Puppy Linux**: This is an all-time favorite of the community. Puppy Linux comes with some programs that are not even included in full distributions.

- **Kali Linux**: Penetration testers and security professionals use Kali Linux for various security testing activities, as it comes with a range of security tools, including supporting live boot.

- **Parrot Security**: This is another operating system, like Kali, that comes with a range of security tools and supports live boot.

Other than live boot-supported distros, there are operating systems that enhance anonymity and are considered secure. As we already discussed virtual machines, you can consider building a virtual machine with a secure operating system:

- **Tails**: Tails is a Debian-based Linux distribution that is designed to provide privacy and anonymity. Connections made outside of it will go through the Tor proxy chain. The Tor Project sponsored Tails. We will discuss the Tor Project and the Tails operating system in *Chapter 10, Proxy Chains and Anonymizers*.

- **Kodachi**: This is another Debian-based distribution designed to provide anonymity and security. Kodachi connects to the outside over secure VPN connections. All network traffic is encrypted, and a free VPN is preconfigured in Kodachi. Also, Kodachi is famous for its anti-forensic operating system, which makes it very hard for memory-related forensic investigations to be carried out. Kodachi supports the DNSCrypt protocol, which encrypts requests sent to OpenDNS to eliminate DNS-level information leakage.

- **Whonix**: Whonix, commonly known as TorBox in the security community, is another Debian-based Linux distribution, designed to ensure privacy and anonymity. Whonix is designed to provide the security of both virtual technology and the Tor chain proxy. Whonix is preconfigured with maximum security settings to be anonymous, and all traffic is redirected through Tor.

- **Subgraph OS**: Subgraph OS is another privacy- and anonymity-oriented operating system that uses the concept of sandboxing. Sandboxing is a technology that uses virtualized or isolated space to execute programs. Custom applications running on Subgraph will be executed in isolated sandboxes to provide security.

- **Qubes OS**: Qubes OS uses separate virtual machines to run each application to provide maximum security. For example, if the user opens a browser, it will be running on a separate virtual machine.

We will discuss these secure operating systems further in *Chapter 10, Proxy Chains and Anonymizers*, where we will be going through the installation, advantages, and disadvantages of each operating system and how we can incorporate secure operating systems with the virtualization and other technologies we discussed. Once we have deployed virtualization and live boot operating systems, that will reduce the footprint while interacting with the internet. Let's look at some commonly used operating systems and how we can improve privacy and anonymity.

How you can improve privacy and anonymity on Windows 10

Windows 10 is one of the most used operating systems. But when it comes to privacy, Windows 10 has a bad reputation. There are ways that we can improve privacy and anonymity:

- Configure the privacy settings in Windows 10 so as not to share information with Microsoft.

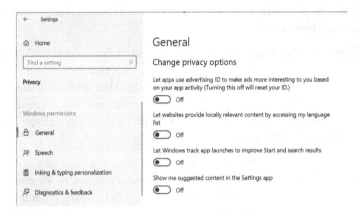

Figure 8.15: Configure the privacy settings so as not to share information with Microsoft

- Disable storing activity history on the device.

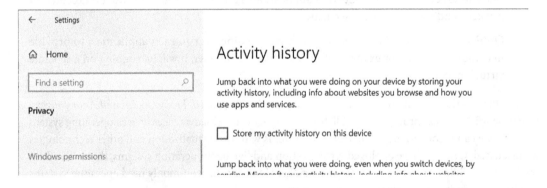

Figure 8.16: Disable storing activity history on the device

- Disable **Getting to know you** on the **Inking & typing personalization** page.

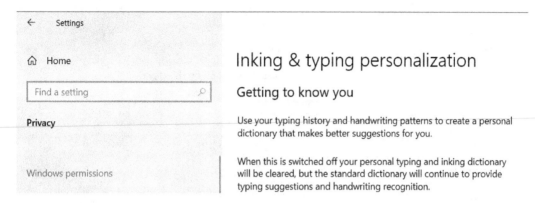

Figure 8.17: Disable Getting to know you

- Enable setting a random hardware address on new networks.

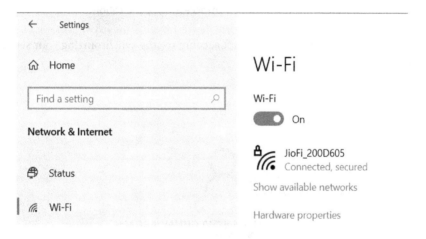

Figure 8.18: Enable random hardware addresses for new networks

- Disable the toggle for applications to access your device location.

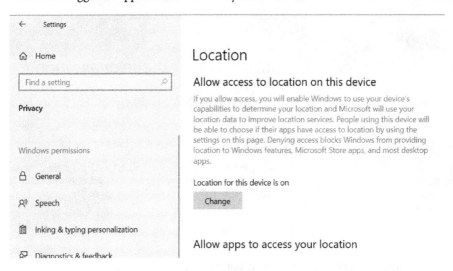

Figure 8.19: Prevent apps from accessing your location

- Switch to a local account from a Microsoft account to stop synchronizing your settings and information with Microsoft.

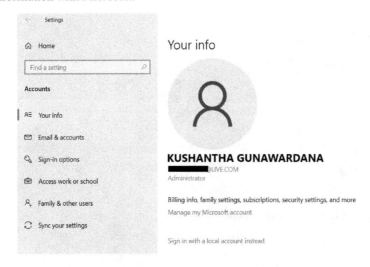

Figure 8.20: Switch to a local account by clicking on the Sign in with a local account instead option

- Disable **Tailored experienc**es. Microsoft will use your data to offer personalized tips, ads, and recommendations if not disabled.

Figure 8.21: Disable personalized experiences

These configurations can provide you with more control over your data on Windows 10 and improve your privacy and anonymity.

How you can improve privacy and anonymity on your Android or iPhone

When you are using mobile devices, you can use similar techniques to what we discussed in *Chapter 7, Introduction to Cyber Anonymity,* to improve security and privacy, including the following:

- Ensure physical security all the time.
- Enable a secure key or password to access the device.
- Encrypt the storage.
- Use a VPN.
- Use secure browsers.
- Use encrypted messaging apps, such as Signal.
- Use secure email services, such as ProtonMail.
- Use secure email clients – ProtonMail also provides a secure email client.
- Use temporary or disposable email addresses when necessary.

- Use temporary or disposable mobile numbers when necessary.

- Configure proper application permissions. For example, don't allow accessing location data for apps that don't require it.

- Disable unwanted connections. If you are not using Bluetooth, turn it off.

- Disable cookies on browsers.

- Disable voice assistance services as if they are kept on, they are listening to you all the time. You can turn off Siri on your iPhone by accessing the settings and tapping on the **Siri and Search** settings. If you are using an Android device, you can go to Google Assistant under **Account Services** and turn it off.

- Stay away from social media. If you use social media, don't configure social media client apps to auto-upload photos taken from your device. But the best way to ensure privacy is to stay off social media.

- Update your device's operating system regularly.

- Use a mobile security solution such as antimalware and antivirus solutions. Many vendors provide free mobile device security solutions, such as Avira, Avast, AVG, and Bitdefender.

When using mobile devices, you need to be mindful of scammers. Some mobile apps, such as Truecaller, can be useful to find out who a caller is when you are receiving a call from an unknown number. Truecaller also offers scam protection on SMS messages and calls. But the catch is, even Truecaller collects information about users. You can use Truecaller in a passive mode, where you will only be using Truecaller to verify unknown numbers.

If you are using an iPhone, you can disable your caller ID when calling someone you don't want to expose your number to. Once you disable caller ID, the receiver will not see your number, which will maintain privacy.

Preparing applications for anonymity

Once the devices are suitably set up to provide privacy and anonymity, we need to prepare applications as applications also collect information about users and create privacy challenges. The method for preparing applications for anonymity depends on the type of application. Some applications provide full functionality of the application without the user having to provide their true identity or even any identity. This mechanism is known as anonymity and pseudonymity. Both anonymity and pseudonymity are important concepts in privacy. Applications that support anonymity do not require any identity information or personally identifiable information to provide functionality. In general, these applications do not require users to register before using the app. Apps that support pseudonymity provide functionality once users produce any username, term, or descriptor that's different from the actual name of the user. In other words, users can use apps that support anonymity without any authentication or verification, whereas apps that support pseudonymity would require a username, screen name, tag name, or something that the user can decide that doesn't expose the user's real personal information. Some apps, such as banking apps, require real identity information to be entered to access the app.

Mainly, we only need to install the required apps on our devices as having more apps increases the chances that our privacy will be compromised. Once you get rid of unwanted apps from your devices, the next step is to disable app tracking on your devices to stop the apps from tracking your activity. Let's look at a few commonly used platforms and apps as examples of how we can block tracking.

How to block app tracking on Android devices

Since smartphones have access to location data and a lot of other information on your device, they often utilize this information for personalized advertising. Android devices are famous for this. You can turn off location tracking to stop unwanted apps from accessing your location data. Some apps, such as store locators, need this access to show you, for example, the nearest store location, but most apps just use it for advertising purposes. You can disable this by accessing the settings on the Android device, then going to **Apps | Apps permissions**. Under app permissions, you can access the location option and disable access to your device location for unwanted apps. Using the same option, you can review other applications and how other applications access the features of the device. You can disable these functions accessed by unwanted apps.

If not, these applications will have access to your calendar, location, call log, storage, and many other functions of the device.

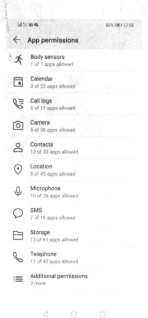

Figure 8.22: Disabling app permissions to device functions

You can disable access to your Android device's features by apps and increase your privacy.

How to block app tracking by iOS devices

Devices with iOS, including iPhones and iPads, track your location and activities. Disabling access to these functions by unwanted apps can increase the privacy of the device. To configure app privacy on an iOS device, you need to access the settings of the iOS device and access the **Privacy** option. When you access the **Privacy** option, you will see how apps have access to many functions on the device.

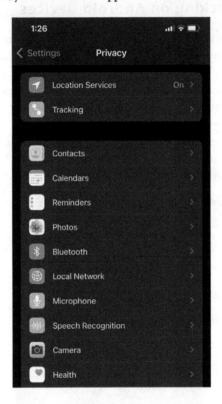

Figure 8.23: App permissions on an iOS device

Once you access the **Privacy** option of the iOS device, you need to access **Location Services** to turn off location services for unwanted apps.

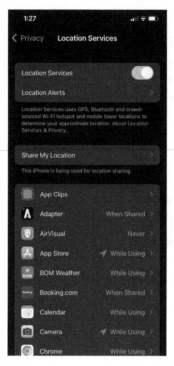

Figure 8.24: Location permissions of an iOS device

You can completely switch off location services for all the apps or you can configure it for each application as required. For example, in the preceding figure, some apps are configured to **Never** for location access and some apps are configured to **While Using** for location access. This will improve the privacy on your device as you are no longer sharing location with all the apps.

Figure 8.25: Tracking permissions

You can also completely disable tracking on your device for apps, or you can configure it so only trustworthy apps track you. If privacy is a concern, then disabling tracking for all apps is the best option.

You can disable location and tracking access for individual apps on a device. But cloud services use a different approach, where they keep all their tracking data on the cloud. Then, you need to access the cloud account and configure the required settings on the cloud account or cloud profile the way you want. For example, if you use Facebook or Google, you need to access your account on the cloud service and disable tracking your actions. If you want to disable this on Facebook, you need to log in to your Facebook account on the Facebook website, go to the **Your Facebook Information** tab under **General account settings**, and select the **Off-Facebook activity** option. Then, you can clear the history and configure the **Off-Facebook Activity for future** option. If you want to configure this on Google, you need to access the **Manage your Google Account** option and access **Data and Personalization** to turn off tracking.

As we discussed in this section, we need to reduce the number of apps installed on our system, as well as configuring app behavior on a local device. If the app is connected to the cloud, often you need to configure tracking from the account section.

Summary

In this chapter, we went through making our presence online anonymous. We learned how to set up cyber anonymity and what is required to maintain cyber anonymity. We also discussed how to become anonymous online under five main topics, as follows:

- Understanding the scope of access
- Planning for connectivity
- Understanding the level of access
- Preparing the device
- Preparing applications for anonymity

First, we tried to understand the scope of access for various apps and services, and then we discussed how to plan connectivity with the objective of maintaining anonymity. We also discussed various levels of access for different resources we use in today's world. Then, we discussed how to prepare a device to be anonymous, including secure operating systems, virtual machines, and live boot systems, which support anonymity. The last topic we discussed was how we can prepare apps to support anonymity.

In the next chapter, we will be looking at another set of interesting topics to discuss how to maintain cyber anonymity and areas and techniques to maintain cyber anonymity.

9

Avoiding Behavior Tracking Applications and Browsers

In the previous chapter, we discussed how to implement cyber anonymity and prerequisites to maintain cyber anonymity. When your objective is to maintain cyber economy, you need to think about multiple layers and plan accordingly. To do that, we need to understand the scope and implement the correct internet connection, as that is the medium to connect to cyberspace. When the medium is secured, you need to prepare a device and applications to maintain cybereconomy, as devices and applications track you in different ways.

This chapter explains how to maintain cyber anonymity and discusses the areas and techniques that we can use to maintain cyber anonymity. As we already know, applications and browsers track our behavior using multiple methods, including cookies and location tracking. When interacting with the internet, we also overexpose ourselves and share sensitive information intentionally or unintentionally, which attackers use to their advantage. We are going to discuss how to maintain cyber anonymity and the areas and techniques that we can use to maintain cyber anonymity in the following topics:

- Avoiding behavior-tracking applications and browsers
- Using secure messaging apps
- Using disposable emails and contact numbers
- Using virtual machines and virtual applications

Avoiding behavior-tracking applications and browsers

In the previous chapter, we discussed how to disable tracking in popular operating systems and some of the most popular apps. Without our knowledge, some apps can still track us, even if we configure cookies to maintain privacy. These apps are programmatically designed to collect information, especially the various web apps we access today. We discussed what web apps can do nowadays in previous chapters. One of the mechanisms that we can use to improve security is anti-tracking browser extensions. There are various anti-tracking browser extensions out there developed by individual developers, companies, and communities. When you access content on a web application or web page, some of the elements on the web page, known as trackers, are designed to collect details about your browsing habits and other information and send it back to third parties. These apps and sites violate privacy and consume lots of resources including your bandwidth to load the pages, as trackers need lots of bandwidth to communicate between the end user devices and the third-party activity tracker. Most web apps and sites collect your information for advertising purposes, but they may have other intentions. Some web apps use this information to provide meaningful content and easy access to information that you are looking for, but some apps can have more dubious intentions.

On the bright side, even though these web apps collect your information, there are various tools and plugins that can be used to make your browsing activity more private. Most of these tools and plugins are designed to let users know what type of information is collected by these apps, and they minimize annoying ads and save resources such as bandwidth from undesired usage.

Tracking is not only active when you are browsing the internet but also when you play games on the internet, watch movies on streaming services such as YouTube, chat with your friends on social media, read emails you have received – especially on free email services such as Gmail and Yahoo – and use apps that are downloaded on app service portals such as Google Play, Microsoft Store, or the iOS App Store. No matter what you do, you will create a digital footprint in cyberspace. Basically, if you have not taken proper precautions, every moment you spend in cyberspace will see you tracked. For example, if you are planning to buy a refrigerator and look at a few options, it's not really a coincidence if you suddenly start getting advertisements for refrigerators on YouTube, news websites, and so on. Mostly, your browser cookies are directly responsible for this, and the rest is done by the trackers. When you access any website, a large number of cookie files are loaded onto your browser for different purposes. Currently, there are a few arguments going on related to allowing users to decide whether they want to be tracked or not by the apps that are installed on their devices. Some larger companies are against this option, as it can prevent the possibility of understanding user behavior and they will not be able to perform targeted advertisements. One such initiative is to introduce an option called **app tracking transparency** (**ATT**) with the iOS 14.5 mobile operating system, which allows users to decide whether they want to be traced or not. This initiative is also challenging to implement, as most apps track you with lines of code and not openly.

So, what's the remedy for tracking? There are a few ways we can protect ourselves from different kinds of tracking, as follows:

- **Configuring your browsers not to track you** – we discussed this in the previous chapters.
- **Using secure browsers** – we discussed secure browsers in previous chapters.
- **Secure extensions to browsers** – installing and configuring secure extensions on browsers can protect you from tracking. As soon as a web application starts tracking, you will be notified. Some browser extensions even block these tracking activities.

- **Desktop applications to protect you from tracking** – there are desktop applications designed to block various types of tracking that come from different sources, and they notify you when doing so.

Since we have already discussed protecting yourself from tracking cookies and using safe browsers, in this chapter we will be focusing on some of the very effective browser extensions and desktop apps that we can use to protect us from tracking. As we know, browsers are client programs that we primarily use to access the internet. Since browsers are the main way to access the internet, if we can block tracking from the browser itself, we should be able to stop tracking at the root.

Browser extensions

Browser extensions can be treated as add-ons, plugins, or additional tools connected directly to a browser. Therefore, we can directly protect ourselves from being tracked by using proper extensions. It's important that we select an extension that is not tracking us, as these extensions can be developed by companies and individuals for different objectives. We are going to go through some of the most popular and trustworthy extensions to see how they work and the advantages and disadvantages of these extensions. Most of these extensions are free, do not utilize a considerable number of resources, and are compatible with most of the popular browsers.

First, we need to know how to install a browser extension. The process is straightforward – you can either search for the desired extension on the web or install it on your browser directly from the extension provider. For example, if you need to install Disconnect, you need to access `https://disconnect.me/disconnect` and hit **Get Disconnect**.

Figure 9.1 – Installation of Disconnect on a browser

When you click on **Get Disconnect**, it will take you to the browser extension installation window. You will see this window when you are trying to install any browser extension in the future. For some browser extensions, you can use a search engine with the extension name. There are also websites with browser extension lists. When you search and click on **Get Extension**, you will be redirected to the browser extension page shown in the following figure.

Figure 9.2 – Installing the extension on the Chrome browser

Clicking on **Add to Chrome** will install the Disconnect browser extension on the Chrome browser. Installation steps are very similar on other browsers. In some browsers, after the installation of the extension, you need to enable it to see it on the browser.

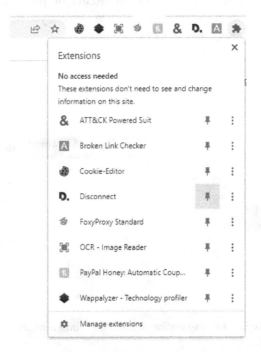

Figure 9.3 – Enabling an extension on the Chrome browser

Once the extension is installed, you might need to enable and pin it to the browser to work. Once pinned, you will see the browser extension on the browser menu. By clicking on the extension icon, you can access the features of the extension.

There are many browser extensions commonly used to prevent tracking and data collection from users by competitors and advertising companies. Some of the browser extensions are not trustworthy, as they also collect information from the users. The following list of extensions is carefully selected based on trustworthiness and functionalities:

- **Disconnect** – Disconnect is a highly recommended browser extension to protect privacy. In 2016, Disconnect was recognized by *The New York Times* as the best privacy tool and one of the top 20 best Chrome extensions by Lifehacker. Disconnect also won an innovation award for best privacy and security tool at South by Southwest in 2015 and was listed as one of the best 100 innovations of the year by *Popular Science*. Disconnect supports a range of popular web browsers. Once you have installed the extension to the browser, when you access any website, it will block trackers and show you what Disconnect has controlled.

Figure 9.4 – A basic version of Disconnect blocks many trackers

As you can see in the preceding figure, Disconnect has blocked 35 trackers, including Facebook, Google, and Twitter-related trackers. Even the basic version of Disconnect can block a range of trackers accessing your information, including Facebook, Google, and Twitter-related trackers. Disconnect also supports major browsers, including Google Chrome, the Samsung browser on mobile phones, Opera, Safari, and Firefox. Once Disconnect blocks different types of trackers and add-ons from websites, they will load much faster than usual, which saves user bandwidth.

Disconnect helps to block a large number of tracking websites, improves website loading time by up to 27%, supports website blacklisting and whitelisting, and has a built-in dashboard that visually displays a range of information, including bandwidth and other information. The only disadvantage is that the basic version has limited capabilities.

When you click on the **Disconnect toolbar** button, it will show you the total number of tracking requests you received on every page you visited in real time. Disconnect will show you in green the number of requests that it blocked. Disconnect will show you in a gray color the requests that are not blocked. You can click a button to see and block or unblock the requests.

In the **Analytics** section, you can view different types of sites that send requests and the status of whether a request is blocked or not. In the **Options** section, you can whitelist or blacklist the site that you are on; this information will be stored in Disconnect and used when you access the site the next time. On the dashboard, you can see the time and bandwidth you saved, and the number of Wi-Fi requests secured by Disconnect if you are connected using a Wi-Fi network.

- **Ghostery** – Ghostery is another safe browser extension that provides a cleaner, safer, and faster web surfing experience. As we discussed earlier, most websites completely rely on tracking and tracker cookies. Some websites even wouldn't work if you block their trackers. For example, when you enable Disconnect and block ads, eBay will not work as usual, as it completely relies on tracker cookies to provide a better experience for its customers. Ghostery offers "smart blocking" that can unblock certain harmless trackers if functionality is reduced when trackers are blocked. It automatically blocks or unblocks accordingly.

  Ghostery is similar to Disconnect; it provides a simple interface but if you need a more detailed view, you have that option too. Ghostery offers a range of security features known as "boosted features" that offer advanced privacy protection, which block trackers, anonymize your data, and block ads and are totally customizable. Ghostery insights provide a state-of-the-art web analytics tool that provides real-time statistics on the performance of every page you visit. The ad-free private search engine provides zero history tracking, zero ads, and no tracing when you search on the web. All these features come as standard with the Ghostery privacy suite.

  You can install the browser extension by visiting `https://www.ghostery.com/ghostery-browser-extension` and selecting the respective browser from the drop-down menu. Currently, Ghostery supports Google Chrome, Mozilla Firefox, Safari, Edge, and Opera browsers. Ghostery's free browser extension provides basic browser protection, whereas Ghostery Plus provides a subscription-based service, which provides basic browser protection and advanced device protection. Once you have installed the Ghostery extension on your browser, you might need to enable and pin it to the browser using the same steps we discussed previously for Disconnect installation. Let's try to access a website and see how Ghostery responds to it.

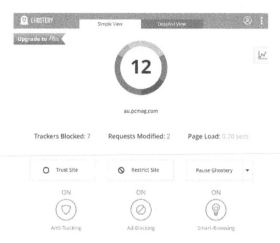

Figure 9.5 – A simple view of Ghostery blocking many trackers

As you can see from *Figure 9.5*, Ghostery has blocked seven trackers, modified two trackers, and loaded a page faster. You are given the option to trust the site (which is similar to whitelisting in Disconnect), restrict the site (which is similar to blacklisting in Disconnect), or pause Ghostery for a desired duration. Also, note that smart browsing is on by default. You can turn it on or off, as well as anti-tracking, ad-blocking, or smart-browsing, from the extension itself quite easily. If you click on the detailed view, you can individually block or allow different trackers. You have the option of restricting the same tracker on all sites or only on the specified site by simply clicking on the detailed view of Ghostery.

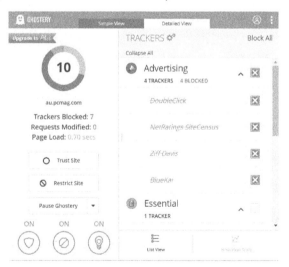

Figure 9.6 – The detailed view of Ghostery provides advanced features

The basic version of Ghostery offers a range of features, but historical stats will only be available once you have upgraded Ghostery to the paid version. The unique set of features offered by Ghostery includes award-winning artificial intelligence-based anti-tracking technology, informing you about what companies are tracking you, and an open source anti-tracking browser extension for a range of popular browsers (as discussed before), increasing the page loading speed by decluttering pages. The easy-to-use interface of Ghostery displays the status of the web page at the bottom of the Ghostery extension, although some websites that use counter-anti-tracking methods might not be identified by Ghostery and might cause Ghostery to not work as expected.

- **uBlock Origin** – uBlock Origin is a wide-spectrum content blocker that optimizes CPU and memory consumption as a primary feature according to the developers. Once you install uBlock Origin, it will automatically enforce the following features:

  - uBlock origin filter lists

  - Easy lists – advertisements

  - EasyPrivacy – tracking

  - Peter Lowe's ad server list (ads and tracking)

  - Online malicious URL block list

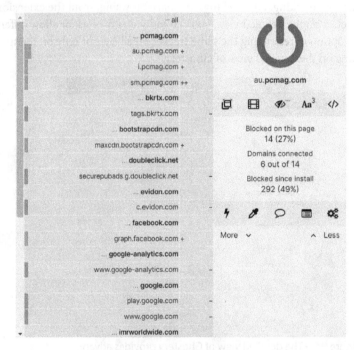

Figure 9.7 – uBlock Origin features

Most importantly, uBlock Origin is free and open source with a public license. Basically, this means people from the tech community volunteer to maintain it and work hard to keep other people safe from ads and tracking. You can install uBlock Origin by visiting `https://ublockorigin.com` and clicking on the **Get uBlock Origin** link. The original browser extension on uBlock Origin was first developed by Raymond Hill back in 2014 to maintain a community-maintained block list and add additional features to improve privacy. uBlock Origin is an open source ad blocker that provides CPU - and memory-efficient technology and supports a range of popular browsers. uBlock Origin has attracted the attention of over 5 million active users of Firefox and over 10 million active users of Google Chrome, winning the prestigious IoT honor of "Pick of the Month" by Mozilla. At the time of writing, uBlock Origin is continuously maintained and under development by the founder and lead developer, Raymond Hill, and remains an industry-leading, open source, cross-platform browser extension that provides privacy from ads and trackers. Apart from being an anti-tracking extension, it also provides a pop-up window blocker, a large media element blocker, and a cosmetic filter disabling feature, and it blocks remote fonts and disables JavaScript on websites.

- **Privacy Badger** – Privacy Badger is another privacy-related browser extension that helps users to stop adversaries and other third-party trackers from tracking and secretly obtaining user behavior information while they are surfing the web. As we discussed earlier, larger companies have the ability to trace your behavior using multiple tracking mechanisms, including third-party cookies. If an advertiser is tracking your activities on multiple websites and pages that you are visiting without your consent, Privacy Badger has the capability to block the advertiser from tracking you. From an advertiser's point of view, it's like you have suddenly disappeared.

Privacy Badger was developed to be used as a single extension to block a range of adversaries and trackers once they have violated user consent. Importantly, Privacy Badger will work without any additional settings or configurations by the end user. It has a set of algorithms that decide whether a website is tracking a user or not. Conventional browser extensions simply block ads while Privacy Badger mainly concentrates on privacy. For example, no ads will be blocked by Privacy Badger if they are not tracking the user. However, trackers will be blocked by Privacy Badger by default. Privacy Badger is purely a tracker blocker.

You can install Privacy Badger by visiting `https://privacybadger.org` and clicking on the respective browser. Privacy Badger supports a range of popular browsers, including Google Chrome, Mozilla Firefox, Opera, and Edge.

Figure 9.8 – Privacy Badger blocks trackers by default

Privacy Badger supports **Global Privacy Control (GPC)** – a specification that provides control to users to notify companies that they would like to opt out of them keeping, sharing, or selling their data. Privacy Badger supports GPC by sending every company that users are interacting with a **do not track (DNT)** signal. When DNT was developed, most companies simply ignored it, so the Privacy Badger browser extension acts as an enforcer of the DNT signal.

Privacy Badger is very easy to configure; it blocks a range of trackers, including invisible trackers. It will not block ads if they are not tracking you, although it is capable of blocking a range of ads. The only drawback of Privacy Badger is that is consumes a noticeable amount of memory.

- **AdGuard** – AdGuard is a collection of open source, free, and shareware products developed by a Moscow-based software development company known as AdGuard Software Limited. One of the AdGuard apps, ApGuard DNS, supports Microsoft Windows, macOS, Linux, iOS, and Android. AdGuard also has a browser extension. It operates with a large number of filters that includes over 5,000 rules, which ensures that no information is collected without user permission. AdGuard blocks ads, trackers, and other types of tools that are designed to collect user data.

Figure 9.9 – The AdGuard range of products to improve privacy

AdGuard has different offerings, including browser extensions to improve privacy. AdGuard has been developed as a desktop application with a range of features, including the following:

- Ad blocking – blocks all kinds of ads.

- Safe browsing – protection against malware. AdGuard checks each web page you surf for malware.

- Privacy protection – AdGuard protects you from all kinds of trackers and analytical systems that have been developed to spy on you.

- Parental control – AdGuard protects children online by removing obscene material and blocking inappropriate websites, and it provides parents with a customizable block list.

- Protect personnel data – AdGuard has a dedicated module developed to protect your data.

- Disguise your presence online – AdGuard supports not only hiding your online presence but also, if required, disguising your online presence and browsing anonymously.

You can download the AdGuard browser extension at `https://adguard.com/en/ adguard-browser-extension/overview.html`. This is AdGuard's lite version that effectively blocks all types of ads and allows safe, fast, and ad-free browsing.

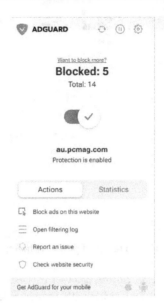

Figure 9.10 – The AdGuard browser extension to block ads

AdGuard provides you with a range of actions through its browser extension to block ads on a specific website straight away, and you can check the security of the website. When you click on the **Check website security** option, it will redirect you to AdGuard's security report of the respective site. AdGuard maintains a database of websites that can produce security reports. A report contains the trustworthiness of the site, child safety settings, and other information related to safe browsing indexes maintained by Google and Yandex. If you go to the statistics, it will show you how many websites were blocked in the past. AdGuard maintains a filtering log that has real-time information about filtering.

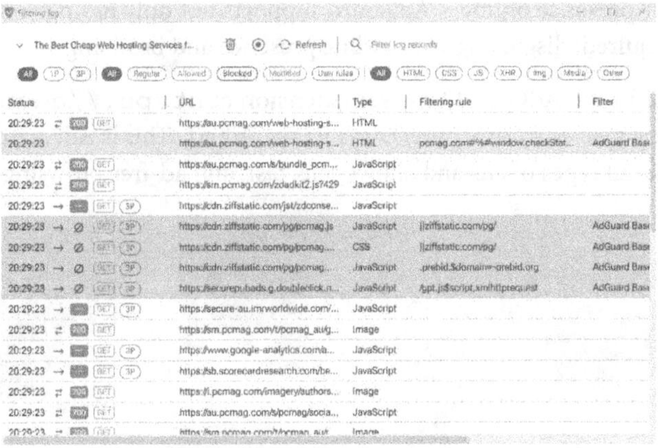

Figure 9.11 – The AdGuard filtering log

When you click on the filtering log, it will show you the real-time status of the filtering, including AdGuard filters and filtering rules. This shows you how many trackers and ads were blocked by AdGuard.

AdGuard also provides a desktop application that you can install on Windows, Android, iOS, or macOS devices to protect you from privacy-related attacks.

Anti-tracking software

In previous topics, we discussed various types of browser extensions that can be used while browsing to block trackers and ads. Sometimes, users might accidentally allow some trackers and ads while browsing. In that case, simple browser extensions might not be effective. Then, another option is to block trackers and ads from other possible layers. Anti-tracking software can be the option for these types of cases:

- **DuckDuckGo** – DuckDuckGo is a popular internet search engine among the security community that emphasizes protecting searchers' privacy and completely avoids filtering the scope of personalized results. DuckDuckGo empowers users to seamlessly control their personal information online without any trade-offs. DuckDuckGo offers a no-cookie incognito mode, which does not allow cookies to track user activities in incognito mode. DuckDuckGo allows users to access the internet without being tracked or traced.

  DuckDuckGo is a search engine and a browser that keeps your data safe and unreachable from advertising companies that use your personal data for advertising purposes. DuckDuckGo also offers a browser extension mode where you can very easily integrate DuckDuckGo into the browser of your choice. When it comes to performance, DuckDuckGo provides faster access, as it disables ads, cookies, and traces and provides complete anonymity. DuckDuckGo comes with an attractive user interface that allows users to access any website without being tracked. As we discussed before, many websites track user data, location, and cookies to collect a lot of information. But with DuckDuckGo, your privacy will be ensured while you are surfing the web. Some users find DuckDuckGo search results are not always accurate, the reason being that DuckDuckGo is not using your private data to refine the results any further.

If you access `https://duckduckgo.com`, you can add DuckDuckGo to your browser as an extension.

Figure 9.12 – The DuckDuckGo search engine and browser extension

Adding DuckDuckGo as a browser extension is straightforward, as we discussed earlier. But if you need to install DuckDuckGo as an app, you need to visit `https://duckduckgo.com/app` and install it on the specific operating system that you are using. DuckDuckGo not only protects your privacy by blocking trackers and ads but also forces websites to establish encrypted connections, when possible, to protect users' data that is communicated between the browser and the server.

- **Cliqz** – Similar to DuckDuckGo, Cliqz is a web browser and a search engine, and provides privacy-oriented services developed by Cliqz GmbH and Hubert Burda Media. Cliqz is available as a desktop application as well as a browser extension. Cliqz provides quick searches with privacy optimization and offers anti-tracking and anti-phishing for a range of operating systems, including Windows, Android, Mac, iOS, and Firefox.

Cliqz works as a first-class doorman, protecting you from unwittingly revealing your data. Just like a doorman, the software only gives access to you, not trackers, ads, and other adversaries. Cliqz comes with anti-tracking, ad-blocker, and anti-phishing capabilities. Cliqz also offers encryption to protect you from sniffing and man-in-the-middle attacks. Cliqz claims that it never collects user data, and its certification company, TUV, has approved Cliqz's architecture in terms of data protection. The Cliqz browser and all integrated function modules are open source. You can download Cliqz by accessing `https://downloads.cliqz.com/CliqzInstaller-en.exe`. It is also available in the Mac, iOS, Android, and Amazon app stores.

- **Brave** – Brave is a free and open source browser that was developed with security in mind. We discussed the Brave browser in a previous chapter.

- **AdGuard desktop app** – We discussed the AdGuard browser extension previously. You can get the AdGuard desktop app by accessing `https://adguard.com/en/welcome.html`.

- **Avast AntiTrack Premium** – You can download Avast AntiTrack Premium by accessing `https://www.avast.com/en-au/antitrack#pc`. Avast AntiTrack Premium blocks trackers from collecting and sharing user data, hides users' online footprints, avoids targeted ads and price manipulation, and provides faster and safer seamless browsing on a range of browsers, including Chrome, Edge, Opera, and Firefox. Avast AntiTrack Premium uses a disguised user profile technique, which masks a user's digital footprint, preventing advertisers from identifying who you are. Avast AntiTrack Premium provides seamless browsing without any disruptions such as breaking web pages and annoying alerts. Some web pages can break when you enable anti-tracking browser extensions. Avast AntiTrack Premium helps to access any web page without breaking and hides your online footprint and browsing history. Avast AntiTrack Premium automatically clears users' histories, cookies, and browsing data.

  Avast AntiTrack Premium was developed initially as an antivirus software that is user-friendly, and it provides an exclusive summary of trackers blocked by the software. Another feature that comes with Avast AntiTrack Premium is an app that supports anonymous browsing, preventing trackers and data collection while surfing the web. This is also capable of effectively blocking a range of scripts that try to collect and track information. The program automatically changes a user's digital footprint to maintain anonymity after deleting tracking attempts.

- **Tor** – Tor is the best of the free browsers and anti-tracking software. It is available on Windows, Mac, and Linux platforms and uses anonymous proxies to maintain a high level of anonymity, which we will be discussing in detail in the next chapter.

Browser extensions and anti-tracking applications can protect websites from tracking and collecting information from you.

When you access web applications and sites, they create cookies in a browser; some cookies are direct cookies, while other cookies are third-party cookies that collect and share information about you with third parties. While avoiding tracking apps and protecting ourselves using browser extensions and anti-tracking tools allows us to surf the web safely, there can still be cookies and scripts created within a browser when you access web applications and websites to collect information about you. Sometimes, attackers and advertising companies can be very tricksy. They might use legitimate websites to store scripts, cookies, and other tracking components within browsers when you access the web. The solution is to use tracking-removal tools to remove cookies and temporary files created when web browsing. There are different kinds of tracking-removal tools available to do this, but some tools can also track you to collect information. Most of these tools are commercial tools, but they also offer free versions with limited features:

- **CCleaner** – CCleaner is one of the best tracking-removal tools available and provides a simple yet effective way of removing cookies, temporary files, recently opened documents, and web history on all web browsers. Even if you use multiple web browsers, CCleaner can remove temporary files created on all browsers in one go. As well as removing the tracking-related files on the system, CCleaner removes unnecessary temporary and log files, which makes a device operate much faster. As well as removing items from your device cookies, junk files, cache, and temporary files, it also helps to remove persistent malware that may have been stored on your system. There are two versions of CCleaner, a free version and a professional version; the free version is good enough for our purposes. You can download CCleaner from `https://www.ccleaner.com/ccleaner/download`.

Figure 9.13 – CCleaner scans and removes temporary files created in browsers

As the preceding figure shows, CCleaner scans all the browsers installed on a system, analyzes the files, and prompts you for action. If you want, you can run the cleaner to clean the files from the system with just one click. It also includes a special browser that can be used without tracking. While the professional version provides a wider range of features, the free version provides interesting features, including a PC health check that automatically analyzes, tunes, and fixes device performance, an app controller that increases the performance of a device, and privacy protection that removes tracking and browsing data.

- **SUPERAntiSpyware** – SUPERAntiSpyware is an AI-powered software that can detect and remove potentially harmful tools, trackers, and software applications from your system, including malware, keyloggers, hijackers, rootkits, Trojans, spyware, adware, ransomware, worms, and **potentially unwanted programs** (PUPs). If you are using Microsoft Defender, which is built in as the antivirus solution in Windows operating systems, SUPERAntiSpyware can boost it. This tool was developed to work with existing antivirus programs, and it will not slow down a system, as it uses a very small amount of your hardware resources.

Figure 9.14 – SUPERAntiSpyware scans a computer for harmful files

You can download SUPERAntiSpyware at https://www.superantispyware.com. SUPERAntiSpyware has a free version and an AI-powered Professional X edition. Even the free version is good enough for tracking removal, but the Professional X version has an attractive set of features not included in the free version, including an AI-backed, real-time scanning engine that is capable of blocking over 1 billion malicious threats, securely deleting malicious files, analyzing a system in depth, cleaning up browser cookies and popups, and stopping ransomware and trackers.

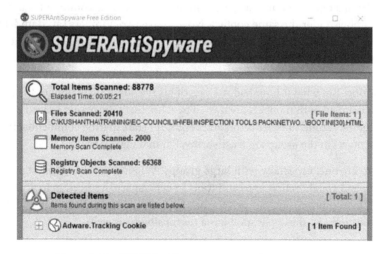

Figure 9.15 – SUPERAntiSpyware scans and removes PUPs

SUPERAntiSpyware can be a very effective tool to remove trackers, including adware and potentially harmful software.

Using secure messaging apps

In the previous topic, we discuss how to avoid browsers and applications from tracking our online behavior. Most browsers and applications track us without our knowledge or consent. Another area to concentrate on when it comes to privacy is communication and collaboration. We typically share private and personal information with our loved ones and people we trust. Communication and collaboration play major roles today, as everyone is connected virtually with each other by various applications, irrespective of geographic location. Traditionally, people met in person to maintain social networks. This was followed by telephone communication.

In today's world, various communication applications and social media are the main ways we maintain social networks. Users typically perform one-to-one communication or one-to-many communications, using various communication tools. These communication tools support video-, audio-, and text-based communication with one recipient at a time, known as a private chat, or with many recipients at a time, called group communication. Typically, these communications can often include personal, private, and confidential information shared with the recipients. Mainly, we need to understand who we are communicating with, especially when communicating in a group.

Sometimes, there can be users in a group that we assume we can trust, but there can be complete strangers in the same group. For example, I have a practice of saving numbers on my phone whenever I receive a call from an unknown number for the first time. I can then call the person back later or identify the same caller next time. Let's say I get a call from someone that was a student when I taught certified ethical hacker training in Australia, and his name is John. I would save his number as "John CEH student Australia." When he phones me the next time, I know that it's him. If the same student phoned me again from a different number – using multiple numbers is common these days – I will save the second number under the same contact. Now, I have two numbers saved under the "John CEH student Australia" contact. Let's say I need to create a group known as "CEH students Australia." Using my chatting application, I will create a group and add contacts to the group. Now, I believe all the contacts in the "CEH students Australia" group are students who were attendees of my certified ethical hacker training. But what if John had use one of his friends' mobile phones to call me in the second instance? I assumed both numbers are owned by John; that's the reason I saved both numbers under his name. But the second number is someone else's. If I start sharing confidential information, assuming all the contacts in the group are trustworthy, I'm making a mistake.

This can happen to anyone, especially with large groups. When it comes to maintaining privacy on communication and collaboration apps and tools in cyberspace, we need to look at these areas. Otherwise, we will be compromising our privacy on these apps. Users have a responsibility when it comes to communication and the application has a responsibility as well.

When dealing with chatting and collaborating apps, we need to look at two aspects:

- One is securing the contacts properly and validating them before we start disseminating information about groups.

- The second aspect is using a secure communication app for communication.

When it comes to communication and collaboration there is a range of apps available that support multiple operating systems, including Windows, Mac, and Linux desktop operating systems and Android, iOS, and Windows mobile operating systems. Messaging apps provide a range of communication and collaboration capabilities, including simple chat, video and audio communication, file sharing, screen sharing, group communication, and location sharing. Typically, messaging apps provide an easy way of maintaining social relationships with colleagues and keeping in touch with family and friends. However, while providing a range of benefits, we need to be aware of online privacy and security concerns related to messaging apps.

When it comes to privacy and security, the potential violations and concerns are as follows:

- Third parties may have access to and can read messages.
- Companies who introduce apps can read and have access to messages.
- Companies who introduce apps can collect information about the users.
- Governments uses apps to collect data about the users and the contents of their messages.
- Interception of communications.
- Access to shared files within the apps by companies and third parties.
- Data collected, including live location information of the users.

To avoid these privacy concerns, we need to evaluate messaging apps before we start using them for communication and collaboration. There are many apps available in all the app stores, including Microsoft Store, Google Play Store, Apple Store, Amazon Store, and desktop app stores, that support communication and collaboration. There are commercial and free apps available for this purpose. Some apps are even open source, developed by communities, and their source code is available to the public. When evaluating these apps for privacy and security, the key considerations should be the following:

- End-to-end encryption – this is when messages can only be read and understood by the sender and the recipient.

- Privacy policy – what does the privacy policy of the app specify it does to protect the privacy of the users?

- Storage encryption – whether the storage of the app is encrypted.

- Status of the source code – is the source code openly available for experts and communities to review to see whether any segments of the code are malicious?

- Use of data – even if the app provides E2E encryption, companies can still collect your data, including locations, duration of communications, device types, and phone numbers.

- Privacy features (self-destruction messages) – users can set a timer for a message to disappear from communication after a preconfigured time. Even when sending an image, users can decide the amount of time it will be displayed before disappearing from a chat.

Encryption

While discussing various ways to maintain anonymity, we looked at how to improve privacy. When it comes to privacy, encryption is a technology or process that we can never exclude. Simply, we cannot talk about privacy in cyberspace without encryption. I'm providing a basic explanation about encryption here to help us understand more about the technologies we can use to improve privacy. Encryption is a process that encodes information so that it can only be accessible by authorized users. Encryption converts human-readable plain text (in encryption, *plain text* refers to information in a raw format or its original representation) to ciphertext (ciphertext is a converted alternative format that cannot be read or understood by humans), which can only be deciphered as plain text by authorized parties. For this purpose, an algorithm is commonly used to cipher and decipher the text. Since everyone uses industry-standard algorithms, users can use a *key* to encrypt the message. If a user needs to decrypt the message, they require the *key*.

There are two versions of encryption:

- Symmetric encryption – when using symmetric encryption, the same key will be used to convert plain text to ciphertext and vice versa. Symmetric encryption is commonly used to encrypt stored data. Typically, the same user will encrypt and decrypt the data. The challenge when using symmetric encryption to secure data in transit is sharing the key because, without it, the recipient will not be able to decipher the data.

- Asymmetric encryption – when using asymmetric encryption, two keys, a "public key" and a "private key," are used to encrypt and decrypt messages. The public key will be distributed among other users while the private key is kept with the original user. When using asymmetric encryption, if one key is used to encrypt a message, the other key of the same key pair can be used to decrypt the message and vice versa. Typically, asymmetric encryption is used to secure data in transit, as symmetric encryption is not practically viable for data in transit, where key sharing between a sender and receiver is a challenge. To implement asymmetric encryption, the keys first must be distributed among the participants for communication. Typically, this is done using digital certificates. *Public key infrastructure* is a term commonly used for an environment that is ready for asymmetric encryption.

When using messaging tools, users don't have to enter keys to encrypt and decrypt data, as it will be managed by the messaging software in the background.

As we discussed, there are many messaging apps available in app stores that support many operating systems, including mobile operating systems such as Android and iOS, and desktop operating systems such as Windows and Mac. Even though all messaging apps provide similar capabilities, some apps have been proven to be more secure than others and provide users with a higher level of privacy. Let's look at some of the popular messaging apps to see the level of privacy that they provide and what the weak messaging apps are when it comes to privacy and anonymity:

- Signal – Signal is a messaging tool that supports cross-platform and E2E encryption for voice and text messages. Signal has a reputation as the most secure messaging app in the industry. It is a free app that is available on a range of platforms, including Windows, Mac, and Linux. As with most messaging tools, Signal requires a phone number to use it. Signal was founded in 2013 but received its highest attention in 2020/2021 when WhatsApp released an updated privacy policy about using user data. Signal is not owned by a large tech company, so they do not rely on ads or targeted marketing. All types of conversations on Signal are, by default, covered by E2E encryption, which means no one, including the company, will have access to conversations except the users who are part of them. Signal also provides self-destructive messages and strives not to collect any information about the user, so the data is only stored on their local device. Due to stronger security, other apps including WhatsApp use Signal's strong messaging protocol for their most secure modes.

- Telegram – This was introduced in 2013 by Russian entrepreneur Pavel Durov and supports multiple operating systems, providing similar types of functionalities as other messenger apps. Telegram provides **E2E** encryption; not even companies, governments, third parties, or hackers have access to messages, only the users in the conversation. This feature is available when Telegram clients call each other, but there isn't a "secret chat" feature. Telegram's extended usage of cloud services to store data means that it stores your chats and files on a secure server, so users can retrieve them easily using any device or operating system. There is an option to "self-destruct" messages when using Telegram; videos and photos can be sent to a recipient with a predefined time to self-destruct once the recipient has received the message. Another favorable feature that Telegrams offers is, once you deactivate your account or are inactive for a certain period, typically 6 months, your account information, including chats and files stored on the secure server, will self-destruct.

- WhatsApp – this app does not require any introduction, as over 1.5 billion users worldwide use WhatsApp. WhatsApp was one of the first apps to introduce E2E encryption to provide privacy, but when WhatsApp was acquired by Facebook, it introduced some level of suspicion regarding the privacy of users. WhatsApp doesn't store information on the WhatsApp servers and allows iOS and Android users to back up messages on the cloud. WhatsApp is owned by Facebook, so data is shared with advertisers for targeted ads.

- Viber – Viber is another popular chatting app operated by a large Japanese company, Ratuken. Users can make phone calls and video calls using Viber. It supports up to 250 users in a group. Viber supports encrypted voice and video calls. Initially, it only supported one-to-one communication encryption, but now it supports encrypted group communication. Viber shows the level of encryption by a color code. When communication is encrypted, Viber shows a green color; a gray color shows communication is encrypted but the contact is not marked as trusted. Red shows failure to authenticate the contact.

- Line – Line is a free communication app, introduced by Japanese internet company Naver for its own staff after the 2004 tsunami. Even though it was introduced to the public in Japan, the later popularity of Line took the app international. Line offers E2E encryption, and users can use the phone number they use to log in to Facebook to register on the app. One of its advantages is that users can communicate without disclosing their contact numbers. That gives users a kind of privacy where they can communicate with people only using their Line ID. In some countries, the Line app provides the capability to find nearby users of the same app with its "Near me" feature, where users can scan and initiate communication with people nearby without disclosing their contact numbers.

- Wire – Wire was introduced by a Switzerland-based company and claims to be a very secure messaging app. Switzerland is generally considered a country with very strict privacy laws. This is one of the reasons that Wire is considered a secure messaging app when it comes to privacy concerns. Wire is also available on Android, iOS, macOS, Windows, and web-based apps, and works on most of the popular browsers. Wire always uses E2E encryption and has an open source messaging app that is available on GitHub to audit if required. Wire only requires an email address for registration; a phone number is not a requirement. Wire also claims that it doesn't collect or sell analytical data of Wire users to third parties.

- Threema – Threema is another messaging app that does not require an email address or phone number to register, which is a big plus point when it comes to anonymity. Threema provides all the standard features of a communication app, including voice, video, and group chat. Threema is not a free tool; it has a one-time fee that users need to pay, and they can then use the app forever. Threema is also an open source communication app that provides E2E encryption so that only the sender and receiver can view messages. It does not require any of your personal data and was developed with protection and privacy in mind. Threema generates very little metadata and is claimed to be totally Swiss-made and **General Data Protection Regulation Act (GDPR)**-compliant. Threema is even available in on-premises mode, so companies can have their own secure communication tool, and nothing will be shared externally.

- Wickr Me – founded by a set of security researchers, Wickr Me is the only tool to provide true anonymity. There are a few versions of Wickr aimed at various audiences, such as as Wickr Pro, Wickr RAM, and Wickr Me. The latter is the app they designed for personal users. Wickr Me doesn't require a phone number or email address to register and does not collect data about users. Wickr Me is more of a collaboration app than a communication tool, as it allows you to share screens and your location with your contacts. Wickr Me provides full E2E encryption for all communications, including videos and pictures. All data is locally encrypted within a device; no one has access to the stored data except the users. Wickr Me offers E2E encryption by default, removes metadata from communication so that files will not transmit with metadata, and has an open source app that offers self-destructing messages. Wickr supports multifactor authentication and does not log or collect user information, including IP addresses.

- Dust – Dust provides users with the opportunity to delete previously sent messages, so messages can be deleted from other mobile phones even after delivering the message. Nothing will permanently be stored in its servers or local devices. They provide E2E encryption for communication. Dust also provides self-destructive messages, allows you to send private messages to your contacts, and you can send messages to a group that are delivered as individual private messages (known as "blasts"). Another interesting feature is that whenever a recipient takes a screenshot using the app, the sender will be notified. In addition to the security and privacy features of Dust, it also offers a privacy protection feature when you surf the web and allows you to implement stealth searching.

- iMessage – iMessage is Apple's proprietary messaging service and is only available on Apple platforms such as iOS, macOS and iPadOS, providing E2E encryption between users. As with WhatsApp, iMessage provides you with an option to back up messages to iCloud. This can be a bit tricky, as iCloud backup uses encryption keys managed by Apple. In other words, anyone who compromises your iCloud could access backed-up iMessage data. iMessage has a few interesting features, such as being able to define how long a picture or video should appear before it self-destructs and the number of times that recipients can view it before it is gone.

These are the most popular messaging apps that are available on the most popular platforms. They have their own pros and cons when it comes to privacy. Depending on your requirements or required level of privacy and anonymity, you can decide which app suits you. Open source apps can typically be trusted, as they maintain a high level of transparency.

Using disposable emails and contact numbers

One of the common ways of exposing ourselves in cyberspace is via emails and contact numbers. For example, there are many requirements to share our email address, contact number, or both to receive activation links, download e-books, access various services, and so on. When we share our email address or contact numbers, companies or attackers use them for various types of malicious purposes. Some companies sell collected information to advertising and marketing companies. Some attackers misuse this information and use it for other attacks.

We need to know the ways to protect ourselves against these attacks without exposing our private email addresses and contact numbers. But this becomes an issue when we need to provide a contact number or email address to gain access to these services. The solution to this is to use virtual numbers and throwaway email addresses. Some services providers such as Yahoo.com provide throwaway email addresses, similar to an alias; once your communication with the service is over, you can disable the email addresses so that they are no longer available for communication.

Another option is using disposable email address. Many disposable email address providers provide disposable email addresses to receive emails temporarily without registration. These email addresses are available either for a certain temporary period or until you refresh the browser. You can just use these email addresses to receive emails or any sort of information, and then you can just forget about them. This is a good way of receiving activation links or download links. Once you have received the link, you can access content without disclosing your real email address, so even if an attacker or company shares or sells this email address, it doesn't matter. However, if you need to reuse the email address, then you need to again register on the services, although most services are available without registration.

The following is a list of disposable email service providers:

- `www.mailinator.com`
- `www.10minutemail.com`
- `www.temp-mail.org`
- `www.guerrillamail.com`
- `www.mohmal.com`
- `www.throwawaymail.com`

If you visit `www.temp-mail.org`, it automatically generates a disposable email address that can be used to receive any activation link or download link.

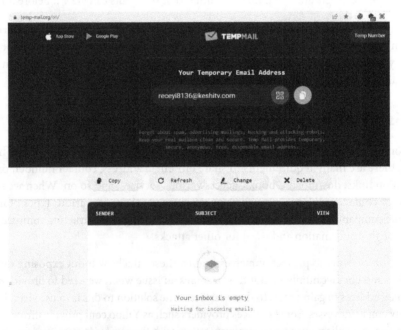

Figure 9.16 – Disposable email address to receive emails

This will prevent attackers from collecting authentic email addresses using different tricks, including providing you with interesting links and registering you on their websites. Once you have used the temporary email address, you can just close the browser.

Keeping your personal contact number safe is extremely important today, as many services use your number to identify and authenticate you. In the past, contact numbers kept on changing. When you moved from one service provider to another, your contact number changed, and when you moved from one city to another, your number changed; it was like your postal address, which changed when you moved from one house to another. But now, in most countries, your contact number is like your passport number or identity number, which never changes. In most countries, you can even change a service provider or phone package without changing the number. Some countries, such as Australia, even have government services that authenticate you by a contact number. When you call them, their systems automatically identify your number, and they can access your information. This makes the contact number more sensitive to privacy breaches than ever before.

If you need to provide a contact number to receive a **one-time password** (**OTP**) or code, then there are multiple options available that allow you to do so without disclosing your real contact number. There are many online and free SMS receiving services to use. First, let's look at free online services that can be accessible without any registration over a browser:

- `https://receive-sms-online.info/`
- `https://receive-smss.com/`
- `https://smsreceivefree.com/`
- `https://pingme.tel/`
- `https://www.freereceivesms.com/en/au/`
- `https://www.receivesms.co/`
- `https://receive-sms.cc/`

All these services provide a similar service; once you access one of the preceding links, you typically need to select the country in which you want to receive the message. For example, receive-SMS-online. info is a free service based on a real SIM and shows the content of the SMS messages you receive to a SIM via a web interface. These numbers are based on real SIMs, but the contents of the messages are shown publicly. If you want to receive a code or OTP, you can use online SMS receiving services. Most of the services do not filter or restrict anything. Since the numbers that these services use are publicly available, some services might have already used them.

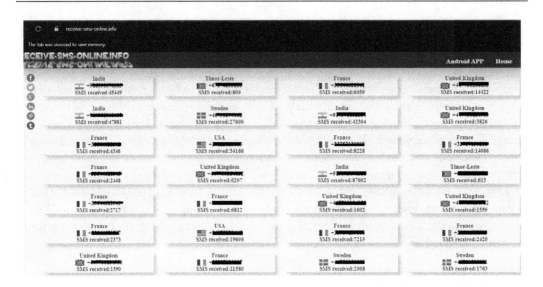

Figure 9.17 – Receiving an SMS for any country

When you access one of the links given previously, you will need to select the country in which you would like to use the number. You can decide based on the country or the number that you would like to use. Once selected, you will be redirected to a web page that receives SMS messages for the selected number.

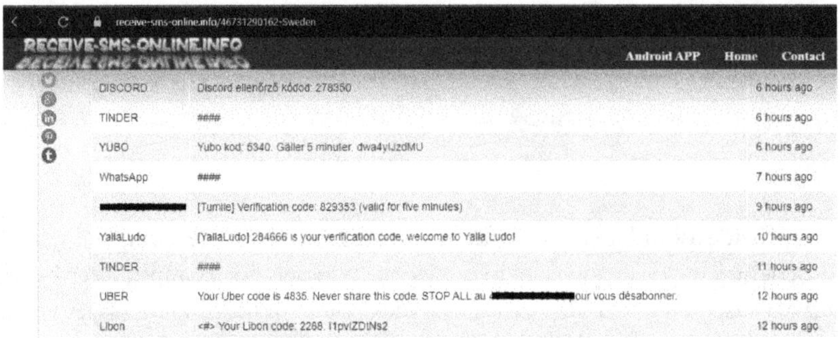

Figure 9.18 – Receiving an SMS for any country

Then, you will see the messages received by the number on the web page; when you use this number on any service, it will display the messages sent to the number publicly. The disadvantage is that all the messages sent to this number will be visible to the public. For some requirements, it's okay to receive messages publicly, as only you can use the code sent to the number for a specific service.

Burner phone

We need a contact number for various reasons, including contacting an emergency service, catching up with family and friends, contacting for official purposes, business requirements, registering with various services, and banking and tax purposes. On top of that, if you want to sell something online or wanted to sell your car, you advertise information with a contact number so that buyers can reach you. When you do this, not only buyers but also various adversaries can get your number easily. Then, you will start receiving promotional calls and text messages, as there are companies that deliberately collect contact numbers for advertising purposes, which is annoying.

Let's say you wanted to register on a dating app and use your contact number. If you use your permanent number, if you change your mind and deactivate your account, the people you connected with on the dating site can still reach you with your number. Sometimes, attackers can even use your number for other attacks. Another problem is that when you use a smartphone, your phone number is integrated with various apps, including messaging apps. Once people get your number, they can search for you on popular apps such as WhatsApp, Line, and Viber. They also can find your profile pictures set up on these messaging apps and other information, depending on the app. This makes it easy for an attacker to gather information about you.

The solution to this issue is a burner phone. Usually, the term *burner phone* refers to a disposable phone with no contract number that is used for a temporary purpose. Typically, it is a prepaid mobile number with a super-cheap handset that you buy on the go, from a shop or supermarket. Some burner phones include a SIM for a certain call time. After the defined period is over, the number will expire. Mostly, these types of packages are used by tourists. For example, if you want to sell something on an online marketplace such as Facebook, you can use a burner phone. Once the item is sold, you can switch off the phone. Since you are using a prepaid number, it will not charge you on monthly basis. Whenever you need, you can top up and start using it. Some packages even support changing numbers. These cheap handsets don't have an option to connect to the internet or install apps. You can usually only call and text using a burner phone. Once done with it, you can get rid of the phone and the number. A burner phone provides stronger privacy as it does not keep any metadata within it. When the phone and SIM are destroyed, there are no traces left.

Virtual numbers

As we discussed, freely available services can be used to receive messages without exposing your number, or you can use burner phones to maintain privacy. But if you don't want others to see the message, you cannot use public SMS receiving services. If you already have an Android or iPhone, you might not want to carry another device, such as a burner phone, with you. Then, the only option available is a virtual number or second number that you can get on your phone.

There are many apps available in app stores that provide a second or temporary number for this purpose. When you download an app, you can obtain a number that can be used as a virtual number. You can use the virtual number to make or receive calls. You can also send or receive SMS messages, depending on the service that you subscribe to. Some services even provide temporary numbers that you can change from time to time:

- Burner – Burner is an app that is available for both Android phones and iPhones but is limited to US and Canadian numbers. It provides a 1-week trial, including 20 minutes of talk time and 60 text messages. If the plan is not upgraded, Burner will be disabled with the number. There are a few subscription plans offered with monthly and annual subscriptions. The Premium subscription offers three numbers with unlimited calls and texts within US and Canada. When you use Burner to call, you will be redirected through Burner, which is known as relaying. When someone calls you, in reality, they are calling Burner; then, they will be automatically relayed to your phone. The Burner app also provides app-level privacy where you can configure a PIN number to use Burner, which can also be integrated with built-in fingerprint or facial recognition on the phone. Burner also provides the option of selecting desired phone numbers from a given list. Using Burner, you can make and receive calls, and send and receive SMS messages.

- Flyp – Flyp is another virtual number app available on both Android and iOS. You can select five numbers based on area codes in the US and Canada. Every number gets unlimited calls and texts. Flyp provides a 7-day trial before you have to upgrade the app. Flyp has a great privacy feature where you can define who can call you back. You can create a list of contacts that can call and text you; calls from other numbers will be automatically blocked.

- 2ndLine – 2ndLine is available for Android phones. Once you install the app, you can select local numbers for the US and Canada with unlimited texts and talk time. Importantly, 2ndLine supports calling and sending text messages from your phone or tablet using your data or Wi-Fi connection. Once you have downloaded the app and picked a phone number, you can access calling and texting features. There is a list of numbers that you can select from. The number from 2ndLine is an additional line to your phone, so you don't have to carry two mobile phones. The paid version provides affordable rates for international calling and texting. You will receive texts to the 2ndLine app, so you can separate personal and other numbers.

- Dingtone – Dingtone is available on both Android and iOS app stores, and you can use it as a second phone number, which provides you with unlimited texts and calling. Number reservation is free, and you can reserve multiple numbers. Dingtone provides business-grade telephone features such ass call forwarding, voicemails, and number blocking. Calling and texts are free within US and Canada and between Dingtone users. Dingtone uses data or Wi-Fi to operate. Dingtone provides affordable international calling rates.

- TextNow – TextNow can be easily used as a second phone number that provides free calling and text messages. TextNow is available in both Android and Apple app stores. TextNow uses data connectivity or Wi-Fi for its service, and you can select a number based on an area code or preferred number. TextNow offers a virtual mobile network service where you will receive a

SIM card for the number, with free calls and texts within the US and Canada. This SIM comes with a data bundle that can be used to connect to the TextNow network. Especially for mobile phones with a dual SIM, this is an impressive option.

- CoverMe – CoverMe is another Android-based app that provides a virtual number to be used as a second number and secure communication over **Voice over Internet Protocol** (**VOIP**) or voice over data networks. CoverMe supports encrypted voice calls, which makes it a very secure app. It supports app-level security by configuring a PIN number. CoverMe works over a data network or Wi-Fi connection. It provides an encrypted secure storage vault on smartphones that cannot be accessed without a PIN number.

- Phoner – Phoner provides a range of additional features, including the ability to mask and hide your caller ID, call forwarding, and voicemails. Phoner is available in Android and iOS app stores and as a web-based application. You can select your second number from a series of randomized numbers. Phoner provides numbers from different countries, but the pricing differs from country to country. Phoner provides an option to send documents as faxes.

- Hushed – Hushed is a very popular virtual phone app that is available on both Android and iOS. Hushed provides virtual numbers for over 40 countries. Hushed uses data connectivity or Wi-Fi networks to operate, so you can start calling from anywhere where data connectivity is available, and it provides a 3-day free trial before you will need to upgrade to a paid plan. Pricing can be different based on the country of the number you selected.

- Google Voice – Google Voice is completely free, but it does not offer complete privacy. The main objective of Google Voice is to provide a second number for call routing. In other words, Google Voice can route calls to all your numbers. You get a permanent secondary number for free that can route calls to all your numbers, including your home number, office number, and mobile number. You can pick up calls from anywhere.

- Line2 – Line2 is designed for small companies or small teams and provides unlimited calls and text messages. Using Line2, you can build a cloud-based telephone solution irrespective of location. Line2 supports both Android and iOS and provides an extension for every team member so that they can call each other and transfer calls.

Using virtual machines and virtual applications

We discussed how we can create a virtual machine with proper network configuration in *Chapter 8, Understanding the Scope of Access*. Once you have created a virtual machine, it can be used for multiple purposes. Remember that virtual machines also work like physical machines on a network. When you check your virtual machine from the network perspective, there is not much difference between the physical machine and the virtual machine. Both will have IP addresses assigned, run an operating system, and have applications installed. When we concentrate on the privacy factor, it provides the advantage that we can revert the virtual machine to a previous state whenever we want, or we can simply reset the virtual machine every time we need to. It will look like we're using a new system every time we access the network.

Once you revert a virtual machine, all the trackers, cookies, and scripts that attackers plant will be removed from the system. From an attacker's perspective, once you revert the virtual machine, your traces will have disappeared. Another advantage of using virtual machines is isolation. Even if your web surfing opens the door for malicious code or malware onto a system, you are still safe. Once you have reverted the virtual machine, they all will be removed. There are various virtualization platforms that you can use to build your virtual machine. It's always a good idea to create a snapshot from the virtual machine so that you can revert to a previous state whenever you want. In *Chapter 8, Understanding the Scope of Access* I explained how to configure a VMware virtual machine.

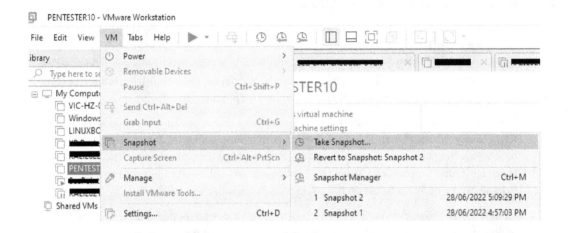

Figure 9.19 – Creating a snapshot of a virtual machine

You can create a snapshot easily on a VMware virtual machine. Just select the virtual machine, go to the **VM** menu, and select the **Take Snapshot…** option. This creates a snapshot of the virtual machine. As you can see in the preceding screenshot, there are two snapshots created.

Whenever you want to revert the virtual machine, you can simply go to the same menu and click on the **Revert to Snapshot** option. Then, the virtual machine will be restored to the previous state, clearing all the traces. This is an advantage of using virtual machines. In simple words, the virtual machine will not create additional protection for anonymity, as on the internet, what others see are packets. It doesn't matter whether the access is virtual or physical in cyberspace. But virtual machines allow you to revert to a previous state, or in other words, every time you access the internet, it's like a brand new entry without any previous states.

We previously discussed preventing tracing and ads at the browser or application level by using anti-tracking applications or browser extensions. Using virtual machines, we can prevent tracing at the operating system level. For example, if you create a snapshot just after the fresh installation of an operating system and then you access the internet, by reverting to a previous state, you can remove all the traces, cookies, temporary files, and caches created within the system within a few seconds. If, next time, you access the internet using the same virtual machine, there won't be any traces of the previous access. In a worst-case scenario, you even have the option to delete the virtual machine completely and recreate another one with minimum effort, unlike rebuilding a physical machine. It's like having a second machine; deleting and creating virtual machines do not have an impact on your daily activities and you won't lose your important data.

As a best practice, what you can do is use your usual computer with browser extensions enabled to access the internet, and then you can clear the artifacts created in your device using CCleaner frequently. If there are any suspicious links or artifacts to check for, you can use your virtual machine.

Once we have taken actions to protect ourselves at the operating system level, then we can think of the application level. As we discussed, applications keep traces of our various activities. When it comes to applications related to privacy, there are two types of apps that we can consider:

- Portable applications
- Virtual desktops and applications

Portable applications

The term *portable applications* refers to applications or software that do not require any installation. In other words, all required files to run the software are typically stored in a single folder or single executable. You can even keep portable applications on a single USB drive, bringing all the required software wherever you go. Previously, you could carry all your software on a CD or DVD; as few people nowadays use CDs or DVDs, you can use a USB instead. Most importantly, you can use the applications and software on any computer and it will feel like it's your computer, as all the required software is present on the USB. When it comes to privacy and anonymity, the best part is that portable software does not store any settings in the Windows registry. Portable software is fully functional like installed software, so we don't have to worry about reduced functionality in portable software.

There are advantages to using portable applications for cyber anonymity, including the following:

- Installation is not required – since installation is not required, the footprint of the application on the device is considerably low. Since the portable applications use a single folder, they are easy to clear.

- No or minimum additional files – typically, portable software doesn't need any additional files, as all the required dependency libraries are bundled with the portable application, except typical runtime files available on the operating system.

- No traces created – portable apps don't keep or create artifacts or footprints on a device; they keep all artifacts and footprints within themselves. When you use the same USB stick with the app on another computer, you will see the artifacts still available on the portable app. If required, simply deleting the portable app or the folder containing the app will remove all the traces and artifacts created by the app, unlike installed apps.

- Easy to clean – since all required portable apps can be stored in a single USB stick, you can bring them wherever you travel, use them on any computer without creating any artifacts on the computer, and destroy all the traces by just deleting or formatting the USB stick properly.

Portable apps have become very popular due to these advantages, and now it's not hard to find portable applications. Many of these portable applications can be located and downloaded for free from `https://portableapps.com/apps`. Portableapps.com offers its own app store, with a collection of apps to select from with proper categories, and you can launch an app straight away without any limitations. Alternatively, some leading software vendors offer portable versions of the same software. It is recommended to download portable apps from trustworthy sources as they may contain bundleware (bundleware is software that comes with additional software as a bundle, which can be malicious or dubious). Downloading from a direct software vendor or portableapps.com is recommended, as portableapps.com scans all the apps available on the site frequently with antivirus engines before publishing them on the site.

These portable apps can save you time while protecting you from traces created while surfing the web. A combination of these tools can be an ideal way of surfing the web. Let's say you want to surf the web without leaving any traces of your presence. You can download the following portable apps to a sanitized USB stick:

- Opera GX portable edition – to access the internet on a faster and more reliable level.

- Eraser – remove and sanitize data without creating traces.

- USB Oblivion – this is an impressive app that removes all traces and evidence that are created in the USB drive that you connect to a Windows device. The reason is whenever you connect a USB to an external drive on a Windows device, it keeps track of the device, including device information. USB Oblivion removes this evidence, including registry entries created on the USB drive.

When you have these three portable apps, you can maintain anonymity while surfing the internet. Let's imagine you have these portable apps on a sanitized USB stick. You can insert the USB stick on the physical machine and connect the drive to your virtual machine. It's easy if you are using the VMware workstation that we created earlier. The only thing you need to do is connect the USB drive to the virtual machine.

Figure 9.20 – Connecting a USB drive to a virtual machine

Once you have connected the USB drive to the virtual machine on the VMware workstation, you can log in to the virtual machine. When you go to File Explorer, you can access the drive. Then, you can access the folder that contains the Opera GX browser or any browser of your choice and open it within the virtual machine. This prevents any artifacts from being created on the virtual machine; even if they are created, we always can revert the virtual machine. Once you complete the work, you can run USB Oblivion to remove the traces created on the USB drive from the virtual machine. Later, if you want to, you can completely wipe the USB drive using Eraser.

There are many interesting portable applications that improve privacy and can be downloaded from portableapps.com, including the following:

- USB Oblivion – this will remove all traces of the USB drive you connected from the device.
- Eraser – the Eraser app can totally shred any data that you want to delete permanently.
- O&O ShutUp – this is a great small tool with which you can configure a range of settings, including tracking Windows 10 in a single window.
- BleachBit – BleachBit is pretty much like other history cleaners but provides more options to remove specific entries across a wide range.
- No Autorun – this tiny portable app prevents you from automatically executing scripts and applications when connecting USB drives to a device.
- fSekrit – fSekrit can save any text file you created encrypted with a password.
- HistoryView/MUICacheView – these portable apps can retrieve browsing history and recent items in a single window.

Virtual desktop and applications

Virtual applications are applications that are optimized to run in a virtualized environment, typically without the requirement to be installed first. An application can reside on the cloud or on-premises but execute on the local device. Using virtual apps provides user privacy and safety, as data will not be stored locally. There are different ways of providing application virtualization. Desktop and application virtualization provides an extra layer of security. In today's world with complex requirements in personal and enterprise systems, virtualization is the best way to handle safety and privacy. On personal and enterprise systems, we have different device form factors such as laptops, desktops, and mobile devices. These devices use different operating systems such as Windows, Linux, Android, and iOS. When you look at the ownership of devices, some are owned by companies whereas others are owned by users. If users access organizational data from their own devices, the privacy and safety of data can be challenging. Even for personal requirements, we might use multiple devices with different ownership to access our data.

Desktop virtualization is the best way to handle this, as it provides access to data using any device type and any operating system, but once you access the data, it will remain in a virtual environment. For example, even if you have access to the data, you will not be able to copy it from the virtual application and paste it into the local device. This separates personal data and organizational data. In other words, desktop and application virtualization provides a range of privacy and security for users, including the following:

- Resource centralization – information and data remain in a single place even though users are given access from anywhere using any device. You can even configure copy-paste protection, where users will not be able to copy organizational data and paste it into personal storage, even if the same device is used.

- Access by any device – users can access resources using any device, including desktops, laptops, or mobile devices, with any operating system. Once you install a remote agent such as Virtual Desktop, you are ready to gain access.

- Policy-based access control – since resources are stored in a central location, strict access control policies can be configured and monitored. You can configure policies or use preconfigured policies to protect your data. Data will be in complete isolation.

- Workspace flexibility – with desktop and application virtualization, users can work from anywhere if they have a decent internet connection.

- Privacy and safety – this completely prevents data loss, as users will not have direct access to data even though they work with it, as the real data is located and processed in a central location. Even external employees and contractors can work with organizational data without violating compliance requirements.

Desktop and application virtualization can even be a solution for individuals. There are different cloud-based service providers that provide desktop virtualization for individuals:

- V2 Cloud – this cloud-based virtual desktop infrastructure provides a virtual desktop as a service. You can create and manage desktops in the cloud efficiently and cost-effectively, and users can access data, business applications, and documents from anywhere and any device without compromising security.

- Shells – Shells offers a personal cloud computer that can start in minutes without compromising your data, even if your personal device is stolen or compromised. You can keep your data on the cloud and Shells provides automatic backups that keep your data safe. Shells also provides E2E encryption. You can get your Shells subscription for under US $5 per month and enjoy a hassle-free desktop experience from any device and operating system of your choice.

- Kamatera – another desktop virtualization solution on which you can decide the size and the operating system. Once you create the cloud PC of your choice, you can start accessing and using it from any device and operating system.

- Amazon WorkSpaces – Amazon WorkSpaces provides another desktop as a service where you can access cloud desktops that manage your data, applications, and documents securely without compromising security.

- Cloud PC – Cloud PC or Windows 365 is a Microsoft solution for desktops as a device in the cloud, but it's still only available as business and enterprise licensing models, even though there is no restriction for individual use.

As well as cloud solutions, there are various on-premises solutions that you can build and use within your infrastructure that provide similar functionalities. As an example, Microsoft Remote Desktop Services provides desktop virtualization with which you can keep data centrally and provide access to users.

When it comes to individual requirements, cloud solutions and on-premises solutions might not be ideal for privacy considerations. The solutions we discussed previously definitely help small, medium, and enterprise-level organizations to protect their data and maintain data privacy and compliance standards, while leveraging the benefits of utilizing the personal devices of users. Companies do not need to invest in devices, as they can use user devices to access virtual desktops and applications while keeping their data protected. That is one of the reasons companies encourage users to use their own devices, known as **bring your own device (BYOD)**. The major concern that many organizations have is data privacy and safety, when users are allowed to use their own devices to access organizational data. Desktop and application virtualization provides the solution for this. Even Microsoft desktop virtualization provides an application virtualization solution where users will not even realize that they are using a virtual application, as it provides the same experience and the same interface that they used to have when using the installed application, yet still provides data privacy and maintains the required compliance standards.

Summary

In this chapter, we discussed different ways and techniques that we can maintain cyber anonymity. We discussed methods to maintain cyber anonymity by using virtual machines at the operating system level. Then, we discussed how we can maintain cyber anonymity at the application level, including application and desktop virtualization. Since we mostly access cyberspace using web browsers, we discussed how to prevent tracking on browsers using browser extensions. Browser extensions can prevent tracking and ads while displaying blocked trackers. There are anti-tracking tools that are capable of blocking trackers at the application level. When the trackers create cookies and scripts for use later or by other applications, we can remove them using trace removal programs. When you go through these methods, you will understand that tracking is possible in multiple layers. When it comes to cyber anonymity, we need to plan and maintain all these layers. Leaving traces on even a single layer will provide enough information to attackers to carry out an attack. We discussed how we can protect ourselves in different layers and maintain cyber anonymity.

This chapter explained how to maintain cyber anonymity and areas and techniques that we can use to maintain cyber anonymity under the following topics:

- Avoiding behavior-tracking applications and browsers
- Using secure messaging apps
- Using disposable emails and contact numbers
- Using virtual machines and virtual applications

By now, you will understand the behavior-tracking techniques and different types of secure messaging apps that can be used to communicate without compromising privacy. Also, you will understand the importance of disposable email addresses and contact numbers. We discussed the importance of virtualization in cyber anonymity and the importance of maintaining cyber anonymity in all the layers.

The next chapter explains the tools and techniques that can be used to maintain cyber anonymity.

10

Proxy Chains and Anonymizers

In the previous chapter, we discussed how to maintain cyber anonymity and techniques that we can use to maintain cyber anonymity related to the following topics:

- Avoiding behavior-tracking applications and browsers
- Using secure messaging apps
- Using disposable emails and contact numbers
- Using virtual machines and virtual applications
- Maintaining cyber anonymity at all the layers

This chapter explains the tools and techniques that can be used to maintain cyber anonymity. During this chapter, you will come to understand what proxy chains and anonymizers are and gain knowledge on censorship circumvention. We will discuss live **Operating Systems** (**OSes**) for maintaining anonymity and how they work, how **Virtual Private Networks** (**VPNs**) work and maintain cyber anonymity, and finally, how we can use logless services to maintain cyber anonymity.

In this chapter, we will discuss the tools and techniques that can be used to maintain cyber anonymity related to the following topics:

- What proxy chains and anonymizers are (for example, Tor)
- Censorship circumvention (for example, Psiphon)
- Live OSes (for example, Tails)
- VPN solutions
- Logless services

What proxy chains and anonymizers are (for example, Tor)

The term **proxy** is used in information technology to define a tool or application that acts as an intermediary between a client and a server. Anonymizers are special kinds of proxy tools that can keep activities on the internet untraceable. In other words, anonymizers are special applications that can make activities that are usually traceable between client and server communication on the internet untraceable. Proxy chains are a series of proxy servers that forward traffic to each other. The objective is to provide higher anonymity than usual proxy servers or anonymizers. For example, even if the traffic is sent through a proxy server, from the internet side, it is always possible to detect the proxy IP. If the proxy servers are accessible and proxy server owners are cooperative, the real IP behind the proxy server can even be discovered by analyzing the proxy server logs. Most anonymizers say that they don't provide information to third parties, but that depends on the requester. When using proxy chains, one proxy always forwards traffic to the other proxy on the chain, making it hard or impossible to trace the real IP of the user.

Anonymizers

The main objective of anonymizers is to provide a reasonable level of confidentiality. Anonymizers support a range of protocols including internet traffic – **Hypertext Transfer Protocol** (**HTTP**) and **File Transfer Protocol** (**FTP**) – and internet services – such as Gopher. Anonymizers typically act as an intermediate application between you and the service that you are accessing. The advantages of using anonymizers include the following:

- **Provide an acceptable level of privacy** – Protects your identity by ensuring your internet navigation activities are untraceable. Most anonymizers protect your privacy, with the exception of intentional disclosure by users – for example, providing personal data in an online survey.

- **Provide an acceptable level of protection against online attacks** – Typically, when users are using anonymizers, their traffic goes through the anonymizer. In other words, the anonymizer will be the front gate for the internet. Even if there is an attack, it will be directed through the anonymizer, as your presence is hidden behind it. Most anonymizers use their own protected DNS servers to manage the traffic.

- **Access restricted content** – In some countries, governments prevent their citizens from accessing certain web content including inappropriate content, websites that publish information that goes against the government, or sensitive information concerning national security. However, anonymizers can be used to access restricted content, as the anonymizers are located outside the country, and their internet connections come through different **Internet Service Providers** (**ISPs**).

- **Evasion of security rules** – Many organizations configure security appliances to block access to certain websites using rules and content filtering mechanisms. Since anonymizers encode and encrypt the traffic within the anonymizer traffic, these appliances will only see the traffic directed to the anonymizers from the client device. However, the reality is that the traffic directed to the anonymizer contains the traffic to restricted destinations. Since the traffic is encoded and encrypted, these appliances, including firewalls, intrusion detection, and prevention systems, cannot block traffic to restricted destinations.

Anonymizers typically access websites for you by protecting your privacy. There are many anonymizers you can use for this purpose including the following:

- **Zendproxy** – Zendproxy is an anonymous web proxy that allows you quick and easy access, protecting your privacy. It changes the IP address connected to your computer quickly and easily. When you surf the internet, your IP determines your location. You don't need to install anything to use Zendproxy – it ports your access over Zendproxy with no hassle. You can control the script on the site and cookies through Zendproxy. There is a range of options that can be configured when surfing the net. You need to enter the URL you desire into the given space and configure whether you want to encrypt the traffic, allow cookies, remove scripts from the website, or remove objects such as ActiveX controls when surfing the internet:

Figure 10.1: Zendproxy provides anonymous access when surfing the internet without any installation

Typically, ActiveX controllers, scripts, and mobile codes are used by web applications to collect information about you. Let's look at the difference between accessing a website directly and through an anonymizer. First, let's access `https://ip.me` directly from the browser:

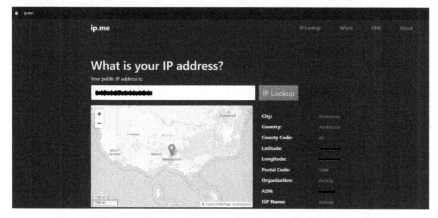

Figure 10.2: https://ip.me shows your location and other information

When you directly access `https://ip.me`, it shows your IP address, location, and GPS information on the site, as the website collects all your information using your IP address. This is how web applications collect your information when you are navigating the internet. When you directly access the internet, it detects your IP address and from that application, detects your location as Melbourne, Australia, your ISP as Belong, the postal code of your location, and GPS information, including latitude and longitude. Scripts running on the web application show your location on a map.

Now, we will try to access the same website using the Zendproxy anonymizer:

Figure 10.3: Accessing through Zendproxy hides your location and other information

As you can see from *Figure 10:3*, when accessing through the Zendproxy anonymizer, it shows a completely different location – the location of the anonymizer in Amsterdam, Netherlands, and the GPS information, postal code, and service provider information are given as LeaseWeb Netherlands. Now, you may notice that the map doesn't appear, as Zendproxy has removed the scripts and objects on the website.

- **Anonymouse** – Anonymouse is another anonymizer that hides your whereabouts when surfing the internet. You can access Anonymouse by accessing `http://anonymouse.org/anonwww.html`. By entering the desired web URL on Anonymouse, you can surf the internet without exposing yourself:

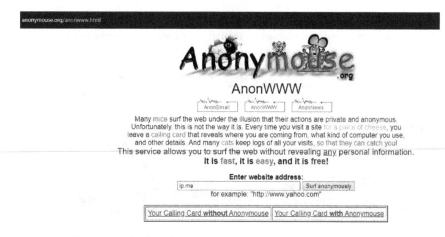

Figure 10.4: Surfing the internet anonymously using Anonymouse

Anonymouse hides all your whereabouts when accessing the internet, providing a faster and easier way of accessing it safely. The best thing is that Anonymouse blocks all the scripts and mobile codes while providing faster access without any ads:

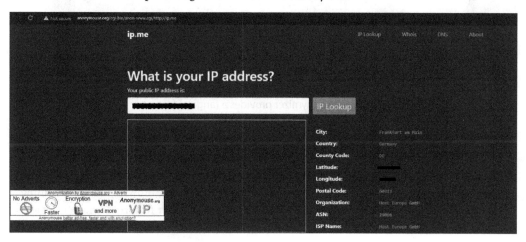

Figure 10.5: Surfing the internet without leaving any traces with Anonymouse

As you can see in *Figure 10:5*, you can access the internet without revealing your real information, the same as Zendproxy. Anonymouse removes the scripts and changes the real location to Frankfurt, Germany, and the other information accordingly.

- **Hide My Ass (HMA)** – HMA provides a range of features in both its free and pro editions. Even the free edition supports a range of security features that users can use even without registering to the site:

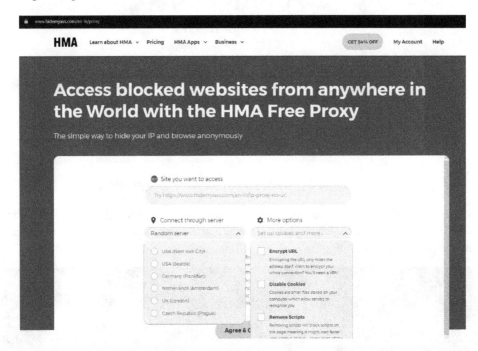

Figure 10.6: HMA anonymizer provides a range of features

As you can see from *Figure 10.6*, HMA provides the option to select a location for users. Even when you are using the free version, you can select one of the desired locations.

Besides this, it also supports typical features such as encryption, disabling cookies, and removing scripts. The HMA free version is good enough for simple browsing and maintaining an acceptable level of anonymity, while HMA pro provides better features for larger devices and even for gamers using gaming apps.

- **Kproxy** – Kproxy is another anonymous proxy available with free and pro versions and provides functionality using a browser extension as well. Kproxy can bypass online content filtering and government- or workplace-related censorship. Even the Kproxy browser extension provides an acceptable level of security, preventing attempts to steal your personal information, including passwords and financial information, even when you are on insecure public Wi-Fi. Kproxy also provides general proxy features such as hiding your IP address to prevent data snooping by ISPs:

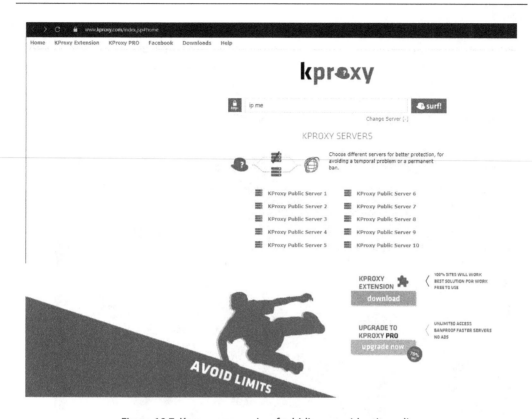

Figure 10.7: Kproxy anonymizer for hiding your identity online

Data snooping by an ISP means that an ISP can sell your browsing activities, behaviors, and history to other parties that want to send you directed ads. Kproxy can prevent this and it also provides the option of selecting a range of Kproxy public servers to redirect your traffic.

- **ProxySite** – ProxySite is a proxy server that provides similar functionality but specializes in social media. If you want to access social media through a proxy, ProxySite is the best proxy server. When accessing any website through ProxySite, it always ensures traffic is encrypted with Secure Socket Layer security even if the site you are trying to access is not a secure site with HTTPS. Users can block ads when using ProxySite or switch between the functions of multiple proxies, allowing users to access sites using multiple countries. Most of the proxy servers for ProxySite are in the United States and the United Kingdom and allow users to remove cookies, scripts, and objects such as mobile code when accessing the web.

Free proxy servers

We learned how to use anonymizers to surf the web without our data being collected while browsing. Many anonymizers we discussed support removing third-party cookies, objects, and scripts while browsing. Many open proxy servers are there if you want to redirect your traffic through proxy servers, but these proxy servers will not remove objects, cookies, or scripts – what they do is redirect traffic through their proxy. There are many sources that provide free proxy lists. If you search for `free proxy server list` on Google, you can find many proxy server lists. If you access `https://geonode.com/free-proxy-list/`, the best thing about Geonode is that it provides users with a long list of proxy servers that can be filtered based on the country, port, anonymity level, such as highly anonymous, anonymous, and non-anonymous, proxy protocol, speed, owned organization, and uptime, among other things:

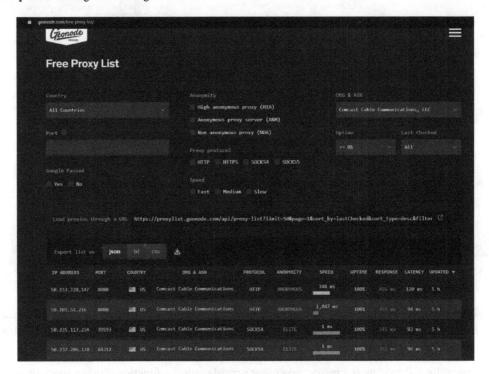

Figure 10.8: Geonode free proxy list

If you need, you can export the current proxy list as a text file, a **Comma-Separated Value (CSV)** file, or a **Java Script Object Notation (JSON)** file. Once you have the information, you can configure your device or the browser to go through the proxy server. Depending on the browser, the configuration is varied. Typically, under the settings, you can find proxy configuration, or if you are using a Windows 10 device, you can go to the device settings, and under **Network and Internet** configurations, you can find the proxy settings. Then, you can configure the proxy IP and the protocol in the proxy settings and save them. Once you have saved the proxy settings, when you browse the internet, your internet traffic will be redirected through the configured proxy server. Similarly, you can find a proxy option on various browsers and different OSes, but the configurations are very similar, as our objective is to set up the IP and the port to redirect traffic:

Figure 10.9: Proxy configuration on a Windows 10 device

You might feel this configuration is cumbersome, as changing the proxy configuration each time is time-consuming. There are tools that can help to switch over multiple proxy servers quickly. FoxyProxy is one of these kinds of tools, which is available as a browser extension that can help you to configure multiple proxy configurations and change quickly whenever you need. You can access `https://getfoxyproxy.org/downloads` to download the FoxyProxy extension, which is available for multiple browsers, including Chrome, Safari, Microsoft Edge, Firefox, and Opera. With the same method we discussed earlier, we can install the FoxyProxy browser extension in the same way and enable it on our browser.

Then, click on the FoxyProxy browser extension and go to the **Options** tab. You can configure the proxy server IP and the port number on FoxyProxy for the desired proxy server configurations. You can select the **Add new proxy** option to add the servers:

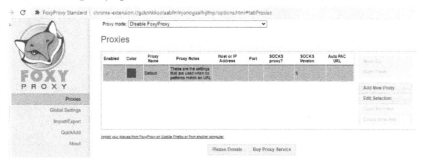

Figure 10.10: Adding proxy servers to FoxyProxy

You can add all your proxy server configurations one by one to FoxyProxy.

Another option on FoxyProxy is configuring an automatic proxy configuration URL. You can use the Geonode free proxy list using the **Load proxies through URL** option given at `https://geonode.com/free-proxy-list/`. You can copy the URL from Geonode and paste the **Automatic proxy configuration URL** configuration into the FoxyProxy **Proxy Details** options. Then, FoxyProxy will use the automatic configuration:

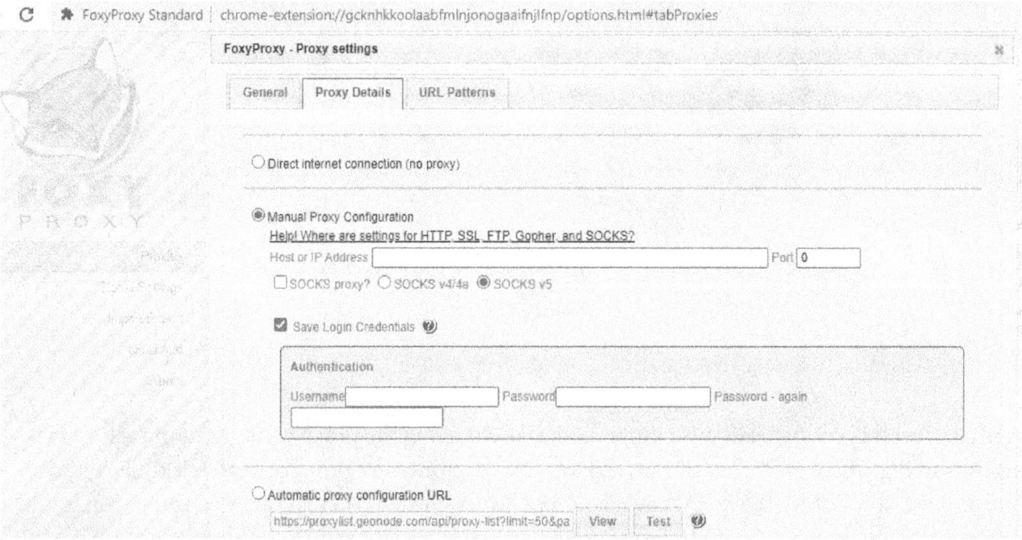

Figure 10.11: Adding a proxy server to the FoxyProxy configuration

Once you have completed adding proxy configurations to FoxyProxy, you can close the **Options** window, go to your browser, click on FoxyProxy, and select the desired proxy server from the list to redirect your traffic. When you are using FoxyProxy, you can configure a list of proxy servers in different countries and different port numbers to redirect your traffic. Whenever you want to redirect traffic, it's just one click away. You can save all the proxy server configurations on FoxyProxy and select which proxy you want to activate:

Figure 10.12: FoxyProxy helps to select the proxy server and redirect the traffic

When you select the proxy that you want to activate from the list of proxy servers you configured, FoxyProxy will start redirecting traffic straight away through the selected proxy server. This is a very efficient way of switching between multiple proxy servers on the go.

Proxy chains

Proxy chaining is a method of forwarding internet traffic through a sequence of proxy servers before traffic reaches its real destination. Once you configure proxy chaining properly and then you try to access the internet, your request first forwards to the first proxy server, and then the first proxy server forwards it to the second proxy server. The second proxy server forwards it to the third proxy server and the third proxy server connects to the destination server. When you send traffic over proxy chains, it will be more secure and increase the level of anonymity, as the destination will only detect the IP address of the first proxy server.

If they need to trace the real IP, an investigation must be conducted to gather information from the log files residing on the first, second, and third proxy servers. If the attacker uses proxy servers from multiple geographies and multiple countries, this process can become very hard or impossible considering the technical, operational, and legal challenges. Proxy chains provide a way of maintaining anonymity, but they don't provide complete anonymity, as every proxy server keeps the information about the connections in its logs. Typically, these log files must be obtained and analyzed to understand the connections made by the proxy servers with each other. When it comes to investigations, this is not an easy task, especially when the proxies are in multiple countries. Owners or administrators of each proxy server should support the process and provide logs if someone wants to investigate the incident. It can be very hard or impossible, but there is a possibility. Many attackers use proxy chains to launch attacks and typically, they use at least five foreign proxy servers when launching attacks:

- **Proxifier** – Proxifier is a powerful tool when it comes to maintaining anonymity, which provides the proxy-chaining capability. Proxifier offers a fully functional copy of the software free for 31 days – then, if you want to continue, you need to purchase the license, which is under 40 USD. Proxifier offers a standard edition and a portable edition (the portable edition does not require installing the software). You can download Proxifier from `https://www.proxifier.com/download/` and configure Proxifier to connect through various proxy servers around the world. You can configure each proxy server IP, port, and socket type in Proxifier. Once the list of proxy servers is configured, the proxy server will forward traffic from one proxy server to another, maintaining a higher level of anonymity. When you configure Proxifier, there are a few things you need to keep in mind:

  - Proxifier supports a range of proxy types, including SOCKS V4, SOCKS V5, and HTTP (SOCKS is an internet protocol that supports exchanging packets within server-client communication through a proxy server. This information can be collected from a site such as Geonode, as we discussed in the previous section).

  - Proxifier supports a mix of different proxy types in a single proxy chain. If there is an HTTP proxy server, it must always be placed as the last proxy server in the proxy chain.

  - When you configure a list of proxy servers for the proxy chain and at least one proxy server is down or not functioning, the entire proxy chain will not function.

  - If the proxy servers connected to the proxy chain have lags (as in, a delay in communication, also known as latency), the total lag will be the sum of all the lags on all proxy servers in the proxy chain.

  - Once all proxy servers are configured to the proxy chain, if the connection to a single proxy server is broken, complete communication to the remote host will be lost.

Once you install Proxifier, you can easily create a proxy chain by clicking on **Proxy Settings** in the **Profile** menu. Then, you can use the **Add** button to add two or more proxy servers. Once the list of proxy servers is added to Proxifier, you can click on the **Create** button to create a proxy chain. Once you create the chain, you can drag and drop previously configured proxy servers into the proxy chain while maintaining the sequence in the way you want:

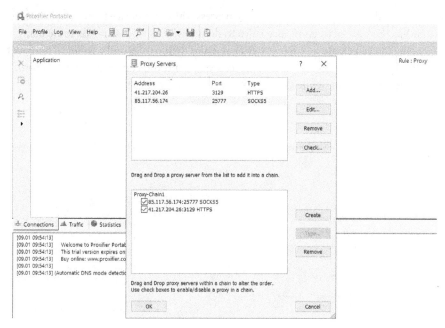

Figure 10.13: Proxifier can be used to create a proxy chain

Once you configure the proxy chain, Proxifier will redirect all your traffic through the proxy chain. The sequence of the proxy servers can always be changed by dragging and dropping them in the way you want. The last proxy server will always be the one connected to the target. You can enable and disable the proxy servers in the list using the checkbox. Proxifier provides all the information related to the connections, traffic, and statistics of the connection:

Figure 10.14: Proxifier redirects all the traffic generated by the device through the chain of proxy servers

Proxifier shows the connections made by each process within the device to the target. Interestingly, Proxifier has the option to configure rules to define which process needs to be using which proxy chain. You can create multiple proxy chains and define rules on Proxifier to redirect traffic according to your requirements – for example, if you want, you can send Google Chrome traffic through one proxy chain, while your email application uses another proxy chain:

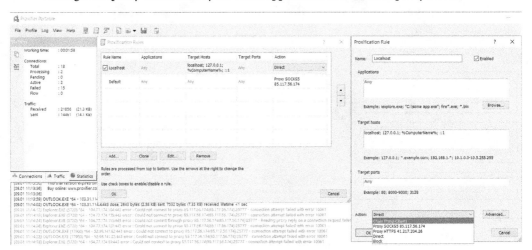

Figure 10.15: Proxifier rules can be configured to direct traffic to different proxy chains

As you can see in *Figure 10.15*, you can configure Proxifier rules to direct traffic through different proxy chains. It shows all the applications are configured to direct traffic through *proxy chain 1*; if you create multiple proxy chains, you can create different rules to direct traffic generated by individual applications through a single proxy or a proxy chain or block traffic. Advanced configuration options are there to redirect traffic through different network interface cards if you have multiple network interface cards connected to your device. It's a great feature to design traffic through different interface cards, especially on a virtual machine with multiple network interface cards.

- **ProxyChains** – ProxyChains is a free and open source program that forces any type of TCP connection initiated or established by any program to go through a sequence of proxy servers. ProxyChains is capable of dynamically forcing any program to go through a chain of proxy servers. ProxyChains hides your IP, sending traffic through restrictive firewalls, and forces a range of programs and utilities, including **Network Mapper** (**Nmap**), Telnet, FTP, APT, VNC, and Wget, to go through proxy chains. For example, if you want, you can use Nmap (one of the free open source utilities most used by security professionals and hackers to scan target devices to collect information, including OS information, open ports, services on the target, and their versions) through ProxyChains. When you use Nmap to scan targets through ProxyChains, from the target side, they will only see the ProxyChains' IP, not the attacker's real IP. ProxyChains has a range of capabilities including the following:

- ProxyChains supports HTTP, SOCKS4, and SOCKS5 proxy servers.

- Supports different types of proxy servers to work on the same proxy chain.

- ProxyChains supports tunneling TCP and DNS traffic through proxy chains.

- ProxyChains hooks traffic generated by network-related functions into dynamically linked programs and forwards it through a chain of proxies.

Being a Unix program, ProxyChins works in the terminal. You can download ProxyChains from GitHub or use Linux security distributions such as Kali Linux or Parrot Security, which have ProxyChains preloaded. If you use a Debian-based Linux distribution, you can get ProxyChains by executing the following command on the terminal:

```
apt-get install proxychains
```

The first thing to do when configuring ProxyChains is edit the configuration file stored in /etc/proxychains.conf. The main configuration is selecting the mode. There are three modes in proxy chains, which you can enable on ProxyChains by uncommenting the desired mode in the configuration file:

- **Strict chain** – All proxies configured in the list will be chained as per the given order.

- **Random chain** – Every connection made will be directed through a randomly selected set of proxies in the list.

- **Dynamic chain** – Dynamic chain will work as per the strict chain, excluding removing functioning proxies from the list.

Then, you can configure the list of proxy servers on the same file using the following format:

```
type host port [username password]
```

Once you have configured /etc/proxychains.conf with the desired proxy servers, it will look as follows:

```
socks5 192.168.67.78 1080 lamer secret
http 192.168.89.3 8080 justu hidden
socks4 192.168.1.49 1080
```

Once you have configured ProxyChains, you are ready to go. When you edit /etc/proxychains.conf with the proxy servers and the mode you desire (dynamic chain is the best mode to maintain anonymity), you can start using proxy chains with any program that you would like to use. It's easy to redirect traffic using ProxyChains – you need to simply follow the following syntax:

```
proxychains [original command]
```

For example, let's say you wanted to surf a website. Then, you need to execute the following command to redirect traffic through ProxyChains:

```
proxychains firefox packtpub.com
```

The following are some of the commands that you can execute to tunnel traffic through ProxyChains:

- `proxychains nmap -sT packtpub.com` – This command will perform a `nmap` scan through ProxyChains.

- `proxychains telnet packtpub.com` – This command will telnet through listed proxy servers.

ProxyChains will tunnel the traffic through the chain of proxy servers that we configured by configuration file and mode.

- **Tor proxy** – Tor is known as the best proxy chaining tool, which has its own browser to simplify the proxy chaining process. When using Tor, we cannot configure the specific proxy servers as with the previously discussed proxy chaining tools, but the Tor browser itself uses its own set of proxies in the Tor network to redirect and tunnel the traffic. Tor supports users reaching blocked destinations and viewing censored content by going through a chain of proxies. Tor distributes your traffic through a network of relays maintained by volunteers around the world. Tor forwards traffic through a worldwide volunteer overlayed network of over 7,000 relays that can conceal the user's usage and location from anyone trying to collect information and traffic analysis. Because of the nature of the Tor network, it makes it very difficult to trace someone's internet activities when they are using Tor. Tor's objective is to protect the privacy of its users and provide freedom and the ability to communicate while maintaining confidentiality and anonymity. You can get your very own copy of Tor at `https://www.torproject.org/download/`. The Tor browser is a completely open source project and is available on Windows, macOS, Linux, and Android.

The name Tor was derived from the privacy project managed by the naval research lab in the United States known as **The Onion Router**. The principle of The Onion Router is to maintain privacy through a number of encrypted layers and encapsulating data like in the layers of an onion. Inside the Tor network, there are a large number of sites known as `<sitename>.onion` that provide hidden services. The design of the Tor network facilitates anonymous web surfing by processing the traffic that goes into or goes through Tor network nodes, immediately relaying to the next destination without collecting or analyzing the destination or original sender. This design obfuscates the original source and the destination of the messages, providing a higher level of anonymity.

The Tor browser provides the easiest method for accessing the Tor distributed network. The installation of the Tor browser is straightforward. Once you download the browser from the original location provided earlier, you can easily install Tor on your device. Then, you can open Tor and access the internet just like any other browser, but Tor ensures anonymity by redirecting traffic through a chain of distributed proxies. The Tor browser has built-in security features to ensure security and privacy, such as disabling JavaScript and other mobile codes, automatic images, videos, and scripts. The Tor browser, including the additional software bundle, is developed and maintained by The Tor Project, which is a non-profit organization continuously carrying out research to improve anonymity in the face of newer developments.

The Tor browser provides access to the internet known as the **searchable internet** or **surface internet** – just like any other browser – but Tor also provides access to the unsearchable internet referred to as the **dark web**. According to research, the dark web has more sites and information than the searchable surface internet, which was primarily only accessible by the Tor browser. When compared to the dark web, the usual surface internet was referred to as *the tip of the iceberg*. Even though some legitimate sites are available on the dark web, including a lot of information related to scientific research, many of the sites on the dark web are said to be illegal sites that promote drugs, counterfeit money, child pornography, weapons, and other kinds of illicit material. Even hackers can be hired on the dark web for hacking assignments. Illegal activities take place because of the higher level of anonymity provided within the dark web.

Apart from the access to the dark side of the internet, Tor also provides many advantages to people who would need to maintain anonymity for ethical reasons. Sometimes, people need to maintain anonymity to prevent discrimination while accessing information as a part of their job. These special categories of users include the following:

- Research students
- Journalists
- Law enforcement bodies
- Military and special agents
- Political activists
- People from countries governed by the repressive regimes
- Users who don't want third parties to monitor their activities online
- People who want to share their genuine thoughts on the internet without discrimination

Tor provides benefits for them by bypassing censorship and avoiding spying and information collection:

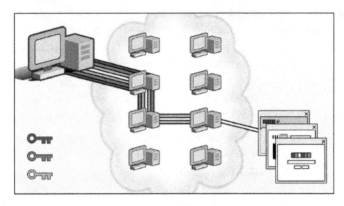

Figure 10.16: Tor tunnels traffic through relays

Tor provides privacy using a network of virtual tunnels on the internet by sending your traffic via three random relays (offered by servers maintained by volunteers around the globe known as **relays**). The last relay forwards traffic to the target via the public internet. As you can see in the figure, when using Tor, it sends the traffic through the computers in the middle and maintains encryption between the client and each relay. As you can see in the figure, it anonymizes the network connection and the IP address that you are connecting from.

However, while Tor was designed to provide security and anonymity, there have been some occasions where authorities revealed the real identities of the attackers even though they were using Tor. Even though the number of occasions where the authorities unmasked the identities is low, this shows that there is a possibility that attackers can be traced. This raises the question of how they can reveal people even if they use Tor when Tor has multiple layers to maintain anonymity. The answer is not hard; Tor provides network-level anonymity, as we discussed before, but there are many other layers still collecting your information, such as the OS and the applications that you use.

While Tor maintains a higher level of anonymity at the network layer, masking your IP, if you are using OSes that collect your information, or if you use your real information on the application, Tor cannot protect you. This is the reason we have discussed multiple layers of anonymity. When you need to maintain a higher level of anonymity, it's essential to maintain anonymity at all the layers. For example, you need to use pseudonyms instead of your real name and shouldn't be revealing your real location, as an IP address is not the only way of revealing your real location.

The reality is that a small clue is more than enough to reveal your entire identity. In particular, investigators can concentrate on behavior-based analysis from open source intelligence and collected information from the searchable internet, the dark web, and various sources that carry telematic information (information that is collected by OSes or applications). This can be used to build a sketch of the picture, adding little pieces of information and clues to it that reveal a real identity, even though this can be a cumbersome process. For this reason, we should maintain anonymity at all layers. Some systems can block Tor. You can use Tor bridges to access Tor in places where it's blocked. In the Tor browser settings, you can configure a Tor bridge by requesting a bridge from The Tor Project, using a known bridge, or building a bridge through systems such as `obfs4`, Snowflake, or `meek azure`.

Censorship circumvention (Psiphon)

In the cyber world, censorship is the suppression of internet activities, including communicating and publishing information that can impact governments and repressive regimes. Then, governments or repressive regimes can force ISPs to enforce censorship. When the ISPs enforce censorship, these websites, services, or applications will not be accessible within these regions.

Internet censorship circumvention defeats censorship using various methods. These methods range from low-tech methods to complex methods depending on the technology that the ISP has to censor, using different services including the following:

- **Using the IP address instead of a URL** – Some censorship can be bypassed by simply using an IP address instead of a real URL (you can find the IP address just by searching the `whois` records or pinging). For example, let's say www.google.com is censored. You can open a terminal and type `ping google.com` to find the IP address and then use an IP address instead of a URL on the browser to access Google:

Figure 10.17: Censorship circumvention using alternative IP

- **URL encoding** – You can bypass censorship using hexadecimal encoding. For example, if you need to bypass censorship enforced on Facebook, we can try hexadecimal encoding. There are many online services for HEX conversion, including `https://www.rapidtables.com/convert/number/ascii-to-hex.html`. Let's go to this link and convert our URL to HEX with a `%` delimiter string, as otherwise, the browser will not accept encoded text. Once the online converter has converted the URL, make sure you enter the `%` delimiter string in front of the hexadecimal output before entering it into the browser:

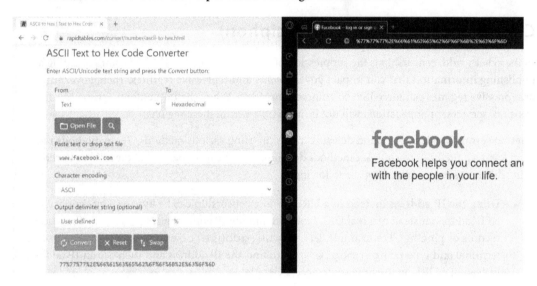

Figure 10.18: Censorship circumvention using hexadecimal encoding

- **Using caches and mirrors** – Even though censorship circumvents the real sources of information, the information may still exist on cached servers or mirrors. There are some resources you can use, including the Google cache and the Wayback Machine. The Google cache is a web cache maintained by Google for all the websites in the world. The Wayback Machine keeps snapshots of all the websites in its repository. You can access these repositories for censored information when the real website is blocked. To access the Google cache, you can use `cache:<website URL>` in Google Search as we discussed before. The Google cache keeps the latest copy of the real website while the Wayback Machine keeps snapshots taken from the website from the day the website appeared on the internet. Let's say we search `cache:www.microsoft.com` on Google:

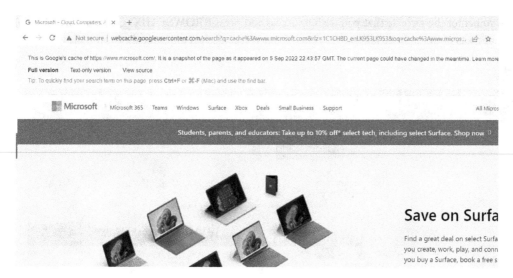

Figure 10.19: Google cache copy of the Microsoft website

As you can see from the preceding figure, Google has cached a copy of the Microsoft website and kept it in Google cache servers. When the next snapshot is taken, Google will replace this copy with the latest snapshot.

If you use the Wayback Machine, it will keep multiple snapshots taken from time to time of the websites. You can access the Wayback Machine at `https://archive.org/web/`. Enter the site you wish to access and click on **BROWSE HISTORY**:

Figure 10.20: The WayBack Machine keeps over 7 billion web pages

Once you enter the website that you wish to access and press **BROWSE HISTORY**, it will show you the snapshots of the website taken over time. Then, you can select any snapshot you wish to view:

Figure 10.21: The Wayback Machine shows the saved snapshots

As you can see in the figure, the Wayback Machine shows all the snapshots taken over time. In this example, www.packtpub.com appeared on the internet for the first time in 2004. When you select 2006, it shows snapshots taken during the year in blue color circles. When you select the blue circle, it will even show you the time of the snapshot. When you select the required time, the Wayback Machine will load the snapshot. This method is useful to access content that is censored:

- **Proxy servers** – We discussed proxy servers in the previous topic and understood how proxy servers can circumvent censorship.

- **VPN solutions** – We will discuss VPN solutions in this chapter's fourth section.

- **Alternative platforms** – The same content can be published in multiple locations by the content publishers. There are decentralized, peer-to-peer platforms supporting sustainable hosting by decentralization. Napster, Gnutella, and ZeroNet are examples of decentralized alternative hosting.

- **Censorship-circumvention networks** – The objective of censorship-circumvention networks is to circumvent internet censorship. As we discussed before, many governments and repressive regimes can enforce internet censorship to prevent people from accessing information. Psiphon, previously known as XP Psiphon, allows users who are using Windows, Mac, or Android devices to connect to censorship-circumvention networks for the purpose of circumventing internet censorship. You can download Psiphon by accessing https://psiphon.ca/en/download.html.

Once you have downloaded Psiphon, you can execute the program straight away without additional configuration and it will connect to a censorship-circumvention network. Additionally, you can select the region to which you want to connect, as Psiphon provides a list of servers from around the world:

Figure 10.22: Psiphon connects to a censorship circumvention network in Singapore

Psiphon is open source software where the source code is available in GitHub but it maintains all servers by itself. When it comes to privacy concerns in Psiphon, it uses cookies and shares access data with Psiphon partners who can see website visits and statistics. According to the privacy bulletin of Psiphon, it occasionally records usage-related data. When it comes to privacy and anonymity considerations, Psiphon may not be the best pick, as the objective of Psiphon is to provide censorship circumvention without drawing unwanted attention, as Psiphon uses common transport protocols. As we discussed, Tor is the best option when it comes to privacy and anonymity, but ISPs can detect if someone uses Tor.

Live OS (for example, Tails)

As we discussed in the previous topic, Tor provides privacy and anonymity at the network level. In other words, maintaining privacy and anonymity cannot be achieved just by using Tor. We need to concentrate on various other layers if privacy and anonymity are our concerns. The main reason is our privacy is compromised at various layers not just at the network layer. The OS in particular collects information, from the time of logging in to the system to an entire set of activities. As a part of the OS, it creates temporary files, a cache, logs, and events for every activity we perform on the device. Depending on the OS, the amount of data collected can vary but typically, any OS which is installed on the device collects information for various reasons, including auditing, troubleshooting, and improving systems performance.

If you need to maintain the highest level of anonymity, you need to get rid of installed OSes and installed applications before you connect to the Tor network. Even though Tor protects you at the network level, your OS and installed application can still unmask you in cyberspace. We already discussed live OSes before and booting your device with a live OS without really installing it on the device. This prevents the creation of logs, events, and temporary files on the device. Especially to overcome the concern that we have with Tor, which is maintaining OS-level anonymity, we can use OSes such as Tails or Qubes OS, which support booting the systems using a USB drive without installing it in the device.

Tails

Tails is a portable OS, fully compatible with running on a USB stick in external hardware, which protects you from censorship and surveillance. Tails is also capable of protecting you from trackers and advertising. Tails on a USB can be used to start your computer instead of the installed OS and doesn't leave any trace once the device is shut down. Tails comes with a bunch of applications that support maintaining privacy and anonymity and are ready to use. Tails is designed with security, privacy, and anonymity in mind – it will be ready to use with even default settings. Tails is a Linux Debian-based, free, and open source OS, which security researchers can always verify and validate its code against privacy and anonymity. Tails provides benefits for a range of users including the following:

- Journalists – When using Tails, journalists can publish their sensitive content while maintaining complete anonymity if they use pseudonyms when publishing content on public forums and access the internet from unsafe locations without revealing their real whereabouts.

- Political activists.

- People from countries governed by the repressive regimes.

- Users who don't want third parties to monitor their activities online.

- Domestic violence survivors – use Tails to escape home violence.

- You – anyone who needs extra privacy in cyberspace.

You can download Tails from `https://tails.boum.org/install/index.en.html` – there are many options when it comes to downloading Tails, including the following:

- USB image of Tails that supports Windows OSes from Windows 7 or later

- USB image of Tails that supports macOS after 10.10 Yosemite or later – supports an Apple M1 chipset

- USB image of Tails that supports any distribution of Linux

- USB image of Tails that supports the command-line terminal for Debian or Ubuntu

- Cloning from an existing copy of Tails on a PC or Mac
- USB image of Tails for USB
- USB image of Tails for DVD
- USB image of Tails that supports virtual machines

Tails doesn't support smartphones or tablets. When downloading Tails, you need to transfer an image to the USB drive. There are many tools that you can use to transfer a downloaded image to a USB drive, but Tails recommends you use `balenaEtcher`, which is a free and open source tool that you can download from `https://www.balena.io/etcher/` depending on your OS. `balenaEtcher` is a cross-platform tool that can flash USB drives and create live SD cards and live USB sticks from `.iso`, `.img`, and even `.zip` folders. If you want to create a Tails virtual machine, you can skip `balenaEtcher`.

If you are using a virtual machine, you can boot Tails directly from the `.iso` file. Once you boot from Tails from the ISO file, it will prompt you with the welcome screen that allows you to select the language and keyboard layout, as with any OS. Then, you need to connect Tails to the network using the network connectivity option –if you are using a virtual machine, Tails will automatically connect using the available network. The next major step is connecting to the Tor network, as Tails traffic is redirected through the Tor network to maintain network-level anonymity:

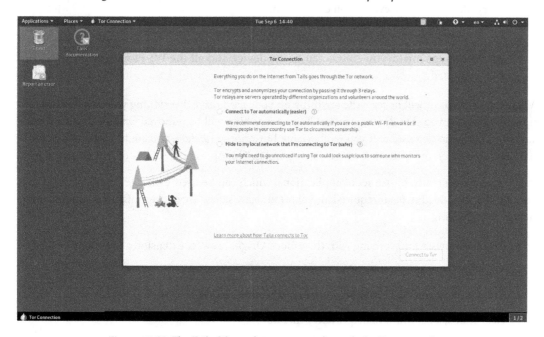

Figure 10.23: The Tails OS needs to connect through the Tor network

Everything you do with the internet from Tails goes through Tor, which maintains network-level anonymity. As we know, Tor encrypts your internet traffic through three relays that are operated by volunteers around the world. There are two options in Tails to connect to Tor:

- **Connect to Tor automatically** – This is the easiest way to connect, especially when you are connecting to the internet using a public internet connection or you need to circumvent censorship.

- **Hide to the local network that I'm connecting to Tor** – This is a safer method to connect to Tor. Especially when your connection is being monitored by an ISP or when you don't want to draw unnecessary attention when using Tor, this is the best option.

If you select the first option, **Connect to Tor automatically (easier)**, Tails will enable another checkbox option, **Configure a Tor bridge**; Tor bridges are secret relays that can evade Tor restrictions. For example, some countries, universities, enterprise networks, or parental controls can block the use of Tor. In these scenarios, you can configure a Tor bridge or Tails will auto-direct to a Tor bridge if your connection is blocked and you need one. Then you can click **Connect Tor** to connect Tails to the Tor network. This will connect Tails to the Tor network and direct internet traffic through Tor.

If you are using any other OS, it will store the following:

- All the websites you visited, including the sites you visited in private mode on the browser

- Every file you opened, even if you delete them

- All the passwords you use, even using a password manager

- All the storage media you connected to the device and all the Wi-Fi networks to which you connected

When you use Tails, it will never write anything on your hard drive. Everything you do will remain in memory. When you shut down or restart Tails, everything will be removed, and Tails will always start from the same clean state. If you really wanted to store some data, you could create encrypted persistent storage on the USB stick where Tails operates. This option is available only if you need it.

Tails comes with a ready-to-use set of applications, which can be used with safe default settings to work on sensitive documents if required and communicate safely. These preconfigured applications include the following:

- The Tor browser for browsing with the uBlock Origin browser extension ad blocker

- LibreOffice, an open source office suite

- Thunderbird, an email client for secure email communication

- KeePassXC, a password manager and generator

- OnionShare, a secure file share over the Tor network

The default settings and the configurations of the applications that are readily available on Tails are safe to use even without additional configuration. For additional security, if any of the applications try to connect to the internet without Tor, the application will be automatically blocked by Tails.

When using Tails, every bit of communication goes through the Tor network. As we discussed before, Tor encrypts the traffic using three secure relays in different locations around the world. Relays are maintained by different volunteers and organizations around the world.

A single relay out of three relays participating in any communication in Tor doesn't know all the information about the communication. When traffic is generated, the first relay only knows about the source of the traffic but not the destination of the traffic. Out of the three relays, the third relay only knows about the destination but not the traffic source. Since the traffic is encrypted using different keys from one relay to another, relays don't really know much information about the communication.

When using Tails connected to Tor, it prevents online censorship and surveillance at multiple layers. This prevents someone from monitoring you and your activities on the internet. Even if you are from a region or country with heavy censorship enforced and even Tor blocked, you can still use Tor bridges to hide that you are connected to Tor. This completely prevents advertisers, trackers, and anyone who is monitoring you from tracing you unless you reveal yourself or your real identity on blogs or any online services. If you are using pseudonyms instead of real identities, online trackers won't be able to trace you again. The best part is using Tails and Tor is completely free!

Qubes OS

Previously, we discussed Tails, which provides privacy and anonymity at the OS and application levels. Since Tails connects to the internet through Tor, it also maintains anonymity at the network level. Qubes OS is a security-oriented OS that takes a different approach to providing security through isolation. Qubes OS is a free and open source distribution that provides desktop computing over virtualization powered by Xen. Qubes OS provides the ability to create and manage isolated compartments known as Qubes, leveraging Xen-based virtualization. Even Edward Snowden, an American former computer intelligence consultant who became popular in 2013 after leaking highly classified information from the American Security Agency, used Qubes OS for his work. You can download Qubes OS at `https://www.qubes-os.org/downloads/`.

As per the Qubes OS tagline itself on the Qubes home page, "*a reasonably secure operating system*," it provides security through virtualization, using compartmentalization and integrated privacy features and allowing freedom. The approach taken by Qubes is virtualization, which provides security-over-virtualization. For example, in Qubes OS, you can run the same browser side by side on different security domains – while logged in to your account on one browser for a specific service, the other browser can work without logging in, as both browsers will run on different virtual instances where it creates two separate sessions on the same browser. The reason is both browsers are working on different domains with separate sessions, technically on separate virtual instances. In Qubes OS, separate virtual machines are used to separate and compartmentalize the environment. Technically speaking, it's like running multiple virtual machines on the same environment, which is integrated by Qubes.

For this reason, Qubes doesn't support the live OS scenario that we discussed before and it would not be totally compatible to install Qubes on a virtual machine, as then it would be nested virtualization.

VPN solutions

When it comes to maintaining anonymity and privacy, we cannot ignore VPN solutions. VPNs have been around for a long time, even before the internet was introduced. Before the internet was introduced, there was no common way to connect multiple networks. The only available option was to use someone else's network that was already implemented to connect two networks. For example, let's say you had a **Local Area Network** (**LAN**) in one geographical location and you had another LAN in a different geographical location. If you needed to communicate from one LAN to another since there was no internet at that time, you could use an already established network maintained by a different company, but the question was that even if the physical network was there, when you connected your LAN to it, the company owning the network could see the communication that you sent through it. To overcome this problem, traffic between one LAN and another LAN was encrypted so that the company providing the infrastructure to connect two LANs could not monitor the traffic. The term "secure VPN" was used for this type of solution.

When the internet was introduced, it provided a great way of not only interconnecting two networks but the same technology also allowing a device in one geographical area to connect to a LAN in a different geographical area through the internet securely. Since the connection was encrypted, they could use even the inherently unsecured public internet for this secure communication. A VPN extends a private network through a public network, which allows users to communicate securely as the VPN provides security, functionality, and management. Initially, VPNs were created by establishing point-to-point connectivity through dedicated circuits managed by different companies, but now, the same implementation can be done over the public internet using tunneling protocols. When using VPN solutions, you can configure different types of VPN solutions including the following:

- **Point-to-site VPN** – When configuring point-to-site VPNs, you can connect one device to a LAN over the public internet securely. This is helpful for working from home and remote worker scenarios.

- **Site-to-site VPN** – This implementation supports connecting two LANs together over the internet or an intermediate connection provided by an ISP. When the connection between two LANs is established, the devices connected to the LANs can communicate with each other as if they are on the same local network.

Typically, in point-to-site VPN implementations, VPN servers are responsible for accepting requests coming from VPN clients. The VPN server has a public IP address to which VPN clients are initiating the connection. Clients usually need VPN client software to be installed on the client device unless the VPN configuration is native (for example, if you use a Microsoft server to configure as a VPN server, Windows clients have VPN client capability built-in and you don't have to install any client software). A VPN client is configured with connectivity information such as the IP of the VPN server, tunneling protocol information, and authentication information.

Even though the classification of VPN solutions is into point-to-point and point-to-site, there are many other classifications of VPNs including the following:

- Based on the tunneling protocol used to establish the VPN

- Based on the topology of the VPN – as an example, a hub and spoke VPN or mesh VPN

- Based on the level of security on the VPN

- Based on the number of simultaneous connections configured on the VPN

When considering privacy and anonymity, this technology provides a range of benefits including the following:

- **Confidentiality** – Since the traffic is encrypted by the tunneling protocols, even if someone sniffs the network, they won't be able to understand the communication without the key.

- **Authenticity** – Since every connection must be authenticated using a password, key, or certificate, this ensures authenticity, which prevents unauthorized access to the communication.

- **The integrity of the communication** – VPN uses hashing functions to ensure the communication is not tampered with.

There are many free and commercial solutions that can be used to maintain privacy and anonymity. We discussed the configuration of a free VPN solution known as OpenVPN in *Chapter 7, Introduction to Cyber Anonymity.*

VPN servers have a capability known as **Authentication, Authorization, and Accounting (AAA)**, which means VPN servers are authenticating the requests, providing access only to the authenticated users, and keeping information about the connections. This means that even though VPN solutions have promising capabilities to improve privacy and anonymity, they also collect and keep information about the connections made. In a corporate environment, it is acceptable to keep a track of all the connections. When it comes to privacy and anonymity, this is a major concern. This capability varies depending on the solutions. Because of this, we need to look at solutions that do not collect or keep information about the connections if we are looking at maintaining anonymity.

As we discussed before, VPN servers or service providers perform authentication. This means they validate the users. Even though some of the VPN service providers claim that they provide complete anonymity, the reality is these service providers can see your IP addresses and your whereabouts, as you need to register on the service by providing your information and they can even keep this data in their logs. Some VPN service providers provide an acceptable level of privacy with strong encryption along with a no-logs policy. They won't guarantee complete anonymity, but they ensure your online activities are kept secret. The best thing about VPN services is that they encrypt your traffic during communication, which enhances privacy even if you are connected to an untrusted public Wi-Fi network or even if your ISP is trying to monitor you. As we discussed before, privacy and anonymity are two different concepts and most VPN service providers use the term "anonymous" just to refer to providing a stronger VPN with zero logs, which actually means they provide privacy. However, the advantage is that even if surveillance agencies wanted to analyze the traffic, they wouldn't be able to do that since the services aren't maintaining any logs. If you want to be truly anonymous, it's better to use a service such as Tor, as we discussed earlier in this chapter, rather than just a VPN solution.

Let's look at some of the popular VPN solutions and evaluate how they improve privacy and anonymity. When comparing available solutions, you will understand which solutions are better in terms of maintaining privacy and anonymity. Since our main objective is to maintain privacy and anonymity, we need to consider whether these VPN solutions provide the following:

- Zero logs or a logless policy – the VPN solution shouldn't be keeping logs about the connections

- Strong encryption with proper implementation – it should provide well-implemented, stronger encryption

- Registration and payment options that encourage anonymity – for example, using pseudonyms to register and cryptocurrency for the payments

- The capability to select connections in multiple geographical areas

- Additional security and privacy features

Considering these facts, let's try to assess some of the popular VPN servers that claim to provide anonymity:

- **VyprVPN** – VyprVPN is a reputed VPN service provider that operates from Switzerland, which we discussed earlier as a highly reputed location when it comes to privacy. VyprVPN maintains a zero-log policy, which means that they don't maintain any logs of the connections made through VyprVPN. VyprVPN also claims that they don't keep any information related to your browsing data and don't pass your information to any third party. This company owns all its servers and hardware, so they do not depend on any other service provider. The VyprVPN no-log policy is audited independently and publicly (they claim that they are the first company in the industry to audit a no-logs policy). You can get VyprVPN at `https://www.vyprvpn.com/buy-vpn` and it provides a 30-day money-back guarantee. VyprVPN confirms that they run their services totally without any third parties and even using their own DNS, known as VyprDNS.

  From the easy-to-manage user interface, you can select any VPN location that you want, with server clusters located in North America, South America, Europe, Asia, Africa, and Oceania from over 700 VPN servers. Each location is powered with top-of-the-line hardware for faster connection and better speeds. They have over 300,000 available IP addresses and over 70 server locations and guarantee that "no third party will ever have access to your privacy."

- **ExpressVPN** – ExpressVPN is another renowned VPN solution that is popular as one of the best VPNs to provide anonymity with its zero-log policy and stealth servers. ExpressVPN provides services from 94 countries at the time of writing and counting. Users can switch often over any country without limitation. ExpressVPN supports users to stream, watch, and listen to censored content and blocked websites around the globe even while traveling. ExpressVPN supports IP address masking, as you are connecting over their servers, which mask your IP address with an ExpressVPN server IP. You can subscribe to ExpressVPN at `https://www.expressvpn.com/order`.

ExpressVPN provides more promising features in terms of anonymity, as it allows users to pay for the subscription using cryptocurrency (for example, you can pay for an ExpressVPN subscription using Bitcoin). ExpressVPN supports cross-platform, as you can use ExpressVPN on Windows, mac, iOS, Android, Linux devices, and even routers and game controllers. ExpressVPN also supports a range of advanced features including the following:

- **VPN split tunneling** – Routing some traffic over a VPN while the rest of the traffic accesses the internet directly.

- **Trusted servers** – These servers do not write anything to hard drives to ensure anonymity.

- **Network Lock kill switch** - This lock keeps data safe by blocking all traffic if the VPN connection is down without directing through direct internet.

- **Private DNS** – ExpressVPN maintains an encrypted DNS service on every server to make sure connections are safe from DNS-level attacks.

- **Best-class encryption** – Data is protected by **Advanced Encryption Standard (AES)** 265, which is the default worldwide accepted encryption standard.

- **Zero-log policy** – ExpressVPN never logs traffic data and connection-related data, DNS queries, or anything that can identify you.

Typically, VPN solutions don't provide complete anonymity, but ExpressVPN provides a faster and more reliable VPN solution that completely conceals your online activities from governments, ISPs, or any surveillance agency trying to track you. While providing a promising set of features, ExpressVPN has the capability of obfuscating your traffic to conceal it from ISPs by disguising it as regular HTTPS traffic.

- **NordVPN** – NordVPN offers secure servers in around 60 countries and over 5,000 secure servers to provide its VPN service. NordVPN accepts cryptocurrency for payment, which ensures anonymity when subscribing to the services. When you subscribe to NordVPN, malware-blocking tools come as part of the subscription. NordVPN maintains a strict no-log policy that does not track, collect, or share your private data. NordVPN's client provides an easy-to-use interface where you can get all these benefits with just one click. You can subscribe to NordVPN at `https://nordvpn.com/pricing/deal-nordvpn/`.

Apart from the preceding services offered by NordVPN, it also provides a set of advanced features including the following:

- **Multiple devices** – You can use up to six devices with a single subscription, securing all your communication, even though you use multiple devices for your communication.

- **Faster VPN** – Typically, VPN services make connection a bit slower, as they do a lot of work in the background, such as encryption and encapsulation, but NordVPN both provides security and speed.

- **Threat protection** – NordVPN supports threat protection against a range of threat actors, including viruses, infected websites, and trackers.

- **Private Internet Access (PIA)** – PIA has gained a good reputation within online communities for its unique advanced privacy and security features. According to their tagline, "*100% risk-free, no logs*," this provides a sense of security for its users. PIA maintains a strict no-logs policy that does not keep any logs for later investigation. PIA provides access for up to 10 devices with a single subscription, offering a dedicated PIA app for every platform. You can subscribe to PIA at `https://www.privateinternetaccess.com/buy-vpn-online`.

Thousands of servers around the world from undefined countries provide easy access to censored content regardless of where you live. PIA is completely open source, which makes it the most secure of all, as anyone can access and check the source code. PIA has built-in ad blocking that blocks ads, trackers, and malicious websites to provide users with a faster and safer internet browsing experience. PIA also supports a kill switch, split tunneling, which we discussed before, torrent support, and highly flexible settings, including support for various themes, OpenVPN encryption, obfuscation, and port forwarding. PIA has also proven its zero-log policy a couple of times even if the origin is the United States – in front of a United States court, it has been proven that PIA doesn't collect any logs of the connections made by the users.

- **PrivateVPN** – PrivateVPN is a relatively new VPN service from Sweden. Especially for new users, it gives a higher level of benefits, as it's just easy to use for anyone, even if you don't have experience using VPN solutions before. You can subscribe to PrivateVPN at `https://privatevpn.com/prices/pvpndealen`. PrivateVPN has a range of additional features to offer, including the following:

- **Zero data logging policy** – No traffic logs or connectivity logs kept.

- **Leak protection** – Your identity will be safe even if you suddenly disconnect from the VPN.

- **Encryption with AES 256** – Highest encryption in the industry.

- **Servers in 63 countries** – Get access from over 60 countries.

- **Kill switch feature** – Automatically suspends internet connection if you disconnect from PrivateVPN.

- **Simultaneous connections** – PrivateVPN is the only provider to connect to six devices simultaneously, all to unique IP addresses.

Logless services

When concentrating on cyber anonymity, it is very important to understand the layers at which your information is collected and how they can be accessed by ISPs, governments, and other snoopers. We have discussed the different layers that collect your information. In the previous section, we discussed how important VPN solutions can be for cyber anonymity, but it can be challenging if snoopers can get information from VPN solution providers, as they can see your activities from the logs that are maintained by them. The best solution for this is logless or zero-log VPN services. The reason is that if the VPN service is not collecting logs, no one, not even VPN solution providers, will not know your activities on the internet. Even in the worst case, if law enforcement wanted to analyze your activities while you were on the VPN connection, they would not be able to do so, as there are no logs whatsoever collected by the VPN solutions provider if they maintain a zero-log policy. Some of the VPN service providers even audit their services on zero-logs policy using independent and public audits. Some VPN service providers have proved during court cases that they never collect logs.

The tricky part is that even though some VPN service providers claim that they don't collect logs, they still keep logs. Because of this reason, before selecting a VPN service provider, you should check whether they conduct independent audits against their zero-log policy, whether they have any mechanism for leak protection, where their services are offered from (as an example, we already discussed Switzerland has strong jurisdiction to protect privacy), and whether they have services such as a kill switch feature to automatically suspend the connection when the VPN is disconnected. Some service providers even keep the services in volatile memory, meaning they never write anything onto hard drives. Importantly, when subscribing to these services, if the users are given the option to make payments using cryptocurrency such as Bitcoin, this will be a definite advantage when maintaining anonymity. Otherwise, users must reveal their real identity to the VPN service provider along with credit card payment information, which will provide the option for them to validate the real whereabouts of the users and compromise anonymity.

We discussed how an OS runs on a USB connected to Tor and can provide anonymity at multiple layers, taking Tails as an example. The idea is to understand how to battle at multiple layers to maintain anonymity in the preying eyes of attackers. There are various options available to ensure anonymity at multiple layers. Apart from the Tor chain proxies, we discussed how VPN solutions can be effective and useful in this regard, especially VPN solutions with zero-log policies. When you look at the multiple layers that we have to consider when maintaining cyber anonymity, you must be thinking it's cumbersome. If a layman or a basic user wanted to maintain anonymity on the internet without much of a technical background, how would they maintain and configure all this before connecting to the internet?

To provide an answer to this question, there are easy-to-use free OSes available for such scenarios that come with secure default configuration, easy-to-use graphical user interfaces, and no additional configuration required to maintain cyber anonymity at multiple layers. The best example is Whonix, which is a security-hardened OS designed to provide anonymity on the internet.

The Whonix OS

Whonix is a security distribution based on Kicksecure. Kicksecure is a security-hardened Debian-based Linux distribution that provides better protection from malware attacks. You can download Whonix, which is an easy-to-use and beginner-friendly open source distribution, from here: `https://www.whonix.org/wiki/Download`.

The main objective of Whonix is to provide stronger privacy and anonymity for users with any knowledge level. In particular, they are trying to keep Whonix an easy-to-use, beginner-friendly, and free OS to overcome all the obstacles we discussed before to being protected and anonymous when surfing the internet. The Whonix website home page provides an interesting comparison between Whonix and VPNs with regard to privacy and anonymity and shows how secure Whonix is when compared to the VPNs that we discussed. Other than hiding your IP address, which is a unique identifier when you access the internet commonly supported by both Whonix and VPNs, Whonix provides a range of safety features in a single OS that most of the VPNs don't provide:

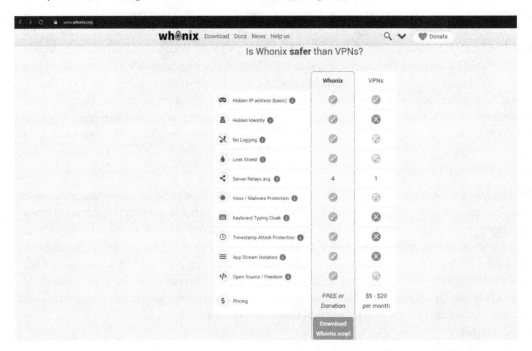

Figure 10.24: Comparison between the Whonix OS and VPN solutions

Whonix supports a range of additional privacy and anonymity features that VPNs don't, including the following:

- **Hidden Identity** – Whonix doesn't just hide your IP address and location; it also protects you from browser and website fingerprinting (protects you from others snooping on your browser activities and your information from being collected on websites).

- **No Logging** – Due to Whonix's fundamental architecture, it cannot log your IP address or activities performed when using Whonix, which makes it the perfect OS for anonymity.

- **Leak Shield** – Whonix can completely protect you from unintentional leaks. Even if you use a VPN, some applications can connect directly to the internet, completely de-anonymizing the user. Whonix has a watertight leak shield, preventing any of the traffic from directly connecting to the internet.

- **Server Relays avg.** – Servers used to relay the traffic must not understand the user or identifiable information about the user; this can only be achieved by using multiple relays maintained by independent owners with onion-layered encryption. Most of the VPN services claim that they provide server relays and then these are typically maintained by the same company. Then, the anonymity is questionable.

- **Virus / Malware Protection** – Even if you are protected at all the layers, if the OS is infected, attackers can still steal your data, including personal information. Whonix is based on Kicksecure, which is a hardened and well-documented open source Linux distribution, so Whonix is inherently secure.

- **Keyboard Typing Cloak** – Nowadays, attackers are very advanced and use complex technologies, including keyboard typing patterns, to understand user behaviors and identities. Whonix protects you from these types of attacks.

- **Timestamp Attack Protection** – Since users connect from different time zones, exposing user time zone synchronization data can also reveal a user's identity. Whonix can protect you from these types of attacks.

- **App Stream Isolation** – There are a set of pre-installed apps in Whonix by default. Whonix has the capability of routing distinct application traffic through a different path, which provides additional security.

- **Open Source / Freedom** – Whonix is a fully open source product, which means anyone can review the code for suspicious content. Independent security experts continuously review the code, which makes Whonix more secure in terms of privacy and anonymity.

- **Pricing** – This is the best part! Whonix is completely free and open source, only relying on donations. Whonix doesn't support any kind of advertising that collects any kind of user data.

- **Kernel Self-Protection** – Whonix has preconfigured kernel hardening settings recommended by the Kernel Self-Protection Project, which makes the Whonix kernel more secure against complex attacks.

- **Built-in advanced firewall** – The Whonix firewall is pre-configured specifically for security and anonymity.
- **Brute force defense** – Whonix provides protection against different types of brute-force attacks using pam tally2, which is a module in Linux that can be configured to lock user accounts after a certain number of failed user login attempts.

Whonix is available on a range of platforms, including Windows, macOS, Linux, Debian, Kicksecure, VirtualBox, Qubes, USB, and KVM.

There are two options to select when downloading Whonix as beginner-friendly:

- Easy-to-use Xface with a graphical user interface
- Terminal only with a CLI for advanced users that only have low-power systems consumption requirements

You can select any of these Whonix versions depending on your requirements. Let's say we've chosen the Windows user version with the GUI Xface beginner-friendly version. This will download the Whonix.ova (**Open Virtual Appliance**) file. Once you have downloaded the .ova file, you can import it to VirtualBox. VirtualBox is a free, general-purpose full virtualization platform powered by Oracle. Installing and configuring VirtualBox is like the VMWare workstation that we discussed before. VirtualBox can be downloaded from https://www.virtualbox.org/wiki/Downloads for free. Let's say you have installed VirtualBox on your device. Then, you can import Whonix. ova directly to VirtualBox using VirtualBox Manager. Open the VirtualBox menu, click on **Import**, select Whonix-XFCE-16.0.5.3.ova, and click **Open**. Once open, you can change the hardware configurations if required, and this will import two virtual systems:

- **Whonix-Gateway** – This connects the traffic to Tor servers.
- **Whonix-Workstation** – This will be in a completely isolated network and connect to Gateway. You will not be able to connect without Gateway:

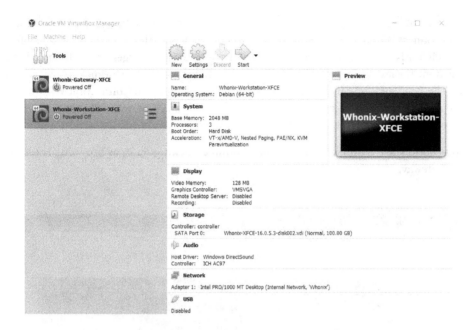

Figure 10.25: Importing Whonix to VirtualBox

First, you need to start Whonix-Gateway, as it will be routing all the traffic to the Tor network. You can start Whonix-Gateway by simply double-clicking on it. When starting Whonix-Gateway for the first time, it will prompt you to configure how you want to connect your Whonix box to the internet. You cannot use Whonix-Gateway to access the internet, as browsing is disabled on Whonix-Gateway by default:

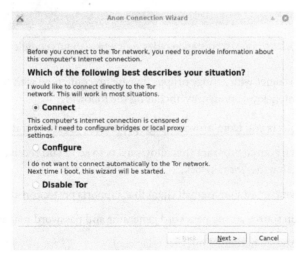

Figure 10.26: Whonix-Gateway settings to connect the Whonix workstation to the Tor network

There are three configuration options as you can see. The first option is the best for most scenarios where it will connect Whonix-Gateway directly to the Tor network. The second option to configure will be useful if Tor is censored and blocked and you need to use Tor bridges, as we discussed in the previous section. The last option is to disable Tor, but it will prompt you next time on how you want to connect. This configuration can be changed later by just running **Anon Connection Wizard** under the **System** tab in the Whonix menu. Once you click **Next**, Whonix-Gateway will connect to the Tor network to provide a higher level of anonymity.

Then, you need to start Whonix-Workstation by double-clicking on it, as browsing is disabled in Whonix-Gateway:

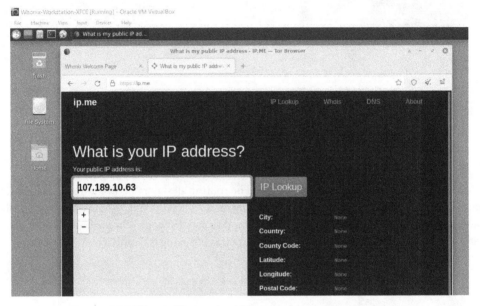

Figure 10.27: Whonix-Workstation hides your whereabouts completely

Whonix-Workstation also comes with a range of apps that are preconfigured with the relevant security settings to ensure application-level anonymity, including the following:

- **Tor browser** – Protects you from browser fingerprinting to get rid of trackers.
- **OnonShare** – Open source program that allows users to send and receive files, maintaining anonymity by utilizing the Tor network.
- **Thunderbird** – Secure open source email client that supports protection and email encryption.
- **KeePassXC** – Open source strong password generator and password manager.
- **HexChat** – Open source hardened internet relay chat.

- **VLC** – VLC media player with JavaScript disabled, protecting from fingerprinting while accessing multimedia files.

- **Electrum** – Electrum is a popular Bitcoin wallet focused on speed and simplicity.

- **Bitcoin** – Decentralized digital currency that enables instant payments over the internet, maintaining anonymity.

- **Monero** – Privacy-centric cryptocurrency, allowing anonymous digital payments without central authority.

These tools are pre-installed in Whonix-Workstation, which is ready to use. When you are using Whonix, it protects you from a range of snooping and information gathering at multiple layers. Whonix is a great, free, open source, less complex solution that can be even used by a beginner without much technical knowledge, as Whonix looks after all the layers that we discussed, maintaining privacy and anonymity with the support of the Tor network.

Summary

In this chapter, we discussed the tools and techniques that can be used to maintain cyber anonymity, using proxy chains and anonymizers and taking Tor as an example, using censorship circumvention, taking Psiphon as an example, using a live OS, with Tails as an example, a range of VPN solutions and how VPN solutions can protect your privacy, and logless services, with Whonix as an example.

After this chapter, readers will understand what proxy chains and anonymizers are and improve their knowledge on censorship circumvention, understand how live OSes maintain anonymity and how they work, understand how VPN techniques work and maintain cyber anonymity, and finally, understand how to use logless services to maintain cyber anonymity.

Cyber anonymity is very essential, as most devices, applications, and browsers collect information on every single activity that you perform in cyberspace. Once you know how to maintain cyber anonymity, depending on what your objective is, it will be helpful to perform tasks without disclosing your information. The objective of this book is to give you enough of an understanding to work with the internet and internet-connected devices safely by maintaining cyber anonymity. This book not only provides conceptual knowledge and understanding but also tools that can be used to maintain cyber anonymity.

Index

A

`Packt.com`

Subscribe to our online digital library for full access to over 7,000 books and videos, as well as industry leading tools to help you plan your personal development and advance your career. For more information, please visit our website.

Why subscribe?

- Spend less time learning and more time coding with practical eBooks and Videos from over 4,000 industry professionals

- Improve your learning with Skill Plans built especially for you

- Get a free eBook or video every month

- Fully searchable for easy access to vital information

- Copy and paste, print, and bookmark content

Did you know that Packt offers eBook versions of every book published, with PDF and ePub files available? You can upgrade to the eBook version at `packt.com` and as a print book customer, you are entitled to a discount on the eBook copy. Get in touch with us at `customercare@packtpub.com` for more details.

At `www.packt.com`, you can also read a collection of free technical articles, sign up for a range of free newsletters, and receive exclusive discounts and offers on Packt books and eBooks.

Other Books You May Enjoy

If you enjoyed this book, you may be interested in these other books by Packt:

Hack the Cybersecurity Interview

Ken Underhill, Christophe Foulon, Tia Hopkins

ISBN: 978-1-80181-663-2

- Understand the most common and important cybersecurity roles
- Focus on interview preparation for key cybersecurity areas
- Identify how to answer important behavioral questions
- Become well versed in the technical side of the interview
- Grasp key cybersecurity role-based questions and their answers
- Develop confidence and handle stress like a pro

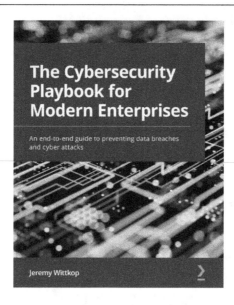

The Cybersecurity Playbook for Modern Enterprises

Jeremy Wittkop

ISBN: 978-1-80324-863-9

- Understand the macro-implications of cyber attacks
- Identify malicious users and prevent harm to your organization
- Find out how ransomware attacks take place
- Work with emerging techniques for improving security profiles
- Explore identity and access management and endpoint security
- Get to grips with building advanced automation models
- Build effective training programs to protect against hacking techniques
- Discover best practices to help you and your family stay safe online

Packt is searching for authors like you

If you're interested in becoming an author for Packt, please visit `authors.packtpub.com` and apply today. We have worked with thousands of developers and tech professionals, just like you, to help them share their insight with the global tech community. You can make a general application, apply for a specific hot topic that we are recruiting an author for, or submit your own idea.

Share Your Thoughts

Now you've finished *An Ethical Guide to Cyber Anonymity*, we'd love to hear your thoughts! Scan the QR code below to go straight to the Amazon review page for this book and share your feedback or leave a review on the site that you purchased it from.

`https://packt.link/r/1801810214`

Your review is important to us and the tech community and will help us make sure we're delivering excellent quality content.

Download a free PDF copy of this book

Thanks for purchasing this book!

Do you like to read on the go but are unable to carry your print books everywhere?

Is your eBook purchase not compatible with the device of your choice?

Don't worry, now with every Packt book you get a DRM-free PDF version of that book at no cost.

Read anywhere, any place, on any device. Search, copy, and paste code from your favorite technical books directly into your application.

The perks don't stop there, you can get exclusive access to discounts, newsletters, and great free content in your inbox daily

Follow these simple steps to get the benefits:

1. Scan the QR code or visit the link below

https://packt.link/free-ebook/978-1-80181-021-0

2. Submit your proof of purchase
3. That's it! We'll send your free PDF and other benefits to your email directly